Technologies for Sustainable Design and Bioregionalist Regeneration

This book explores the role of technology in ecological urban design and regeneration. Part I provides theoretical and methodological insights into technological approaches that offer optimum respect to bioregional, cultural and natural environments, while providing minimum impact and carbon footprint. Parts II and III show contextualized examples that demonstrate the use, or proposal, of sustainable technologies and solutions for regenerating parts of the urban and peri-urban. The case studies present insights from the Mediterranean and the Middle East in a diverse range of spaces, from central urban squares, oblique cities, urban waterfronts, and decaying suburbs to peri-urban areas such as touristic waterfronts, former industrial zones, hyper-commercial areas, wetlands, and parks.

Dora Francese is a Professor in the School of Architecture at the University of Naples, Italy.

Routledge Studies in Urbanism and the City

This series offers a forum for original and innovative research that engages with key debates and concepts in the field. Titles within the series range from empirical investigations to theoretical engagements, offering international perspectives and multidisciplinary dialogues across the social sciences and humanities, from urban studies, planning, geography, geohumanities, sociology, politics, the arts, cultural studies, philosophy and literature.

Published:
The Urban Political Economy and Ecology of Automobility
Driving cities, driving inequality, driving politics
Edited by Alan Walks

Cities and Inequalities in a Global and Neoliberal World
Edited by Faranak Miraftab, David Wilson and Ken E. Salo

Beyond the Networked City
Infrastructure reconfigurations and urban change in the North and South
Edited by Olivier Coutard and Jonathan Rutherford

Technologies for Sustainable Urban Design and Bioregionalist Regeneration
Dora Francese

Forthcoming:
Mega-Urbanization in the Global South
Fast cities and new urban utopias of the postcolonial state
Edited by Ayona Datta and Abdul Shaban

Technologies for Sustainable Urban Design and Bioregionalist Regeneration

Dora Francese

Routledge
Taylor & Francis Group

LONDON AND NEW YORK

First published 2016
by Routledge
2 Park Square, Milton Park, Abingdon, Oxon OX14 4RN

and by Routledge
711 Third Avenue, New York, NY 10017

First issued in paperback 2018

Routledge is an imprint of the Taylor & Francis Group, an informa business

British Library Cataloguing in Publication Data
A catalogue record for this book is available from the British Library

Library of Congress Cataloging in Publication Data
Francese, Dora, author.
Technologies for sustainable urban design and bioregionalist regeneration / Dora Francese.
    pages cm. -- (Routledge studies in urbanism and the city)
    1. Sustainable urban development. 2. Sustainable design. I. United States. Secret service II. Title.
    HT241.F725 2016
    338.9'27--dc23
                                                                    2015034088

ISBN 13: 978-1-138-54718-6 (pbk)
ISBN 13: 978-1-138-99939-8 (hbk)

Typeset in Times
by Sunrise Setting Ltd, Paignton, UK

Contents

Figures

Foreword

COMUNE DI **NAPOLI**

Il Sindaco di Napoli
Municipality of Naples
The Mayor of Naples

What is a cynic?
A man who knows
the price of everything,
and the value of nothing.

(Oscar Wilde)

Providing value to spaces. I welcome Dora Francese's precious work, not only as a relevant scientific contribution to the debate about sustainable urban design but as a practical handbook. It is also important for people such as me, a politician who has to administer the destiny of a big city, conscious of the social and architectural transformations that invest the contemporary metropolis.

Dora Francese's contribution is even more opportune, as it falls into the proximity of the metropolitan city's birth, the new constitutional entity that substitutes the province, and which obliges us "to think metropolitan." Naples, today, is more and more a metropolis, chiefly due to its administrative geography. A metropolis that is aimed at playing a global role, otherwise it will surrender, without lodging an appeal. We are ready for the challenge thanks to this important contribution that you now have in your hands, and also determined to transform these institutional innovations into positives for the whole Campania Region.

My exhortative language is not casual. Urban geography, from the seminal works of Sakia Sassen, teaches us that global competition no longer occurs between the old, spoiled, nineteenth century nation states, absorbed by transnational organizations such as the UN or the European Union. It is ready to be switched on among the metropolitan areas: complex and polyfunctional urban systems, which compete at global scale to catch the best students—thanks to their prestigious universities—and to hold the richest customers—due to their hyper-efficient administrative facilities, reasonable taxation—and finally in order to locate innovative industries—thanks to a net of social capital and high profile services.

Global competition also occurs within the same country, where the municipal federalism logic underlines the chance for the citizen to select the more appealing city, according to his preferences: the metaphor is that of the citizen-consumer. Of course, I am very critical as far as this competition ideology among territories is concerned. I found the under-culture of total competition, to which this rhetoric refers, submerged by social Darwinism. It stresses too much on efficiency and too little on equity, the hazard being that of creating whole cities of a B series, new ghettos for that *lumpenproletariat*, which cannot win *the struggle of the fittest*, to which this vision is inspired. However, willing or nonwilling, this is the scenery in which we should operate: this is the race Naples has to play.

Naples should remain faithful to herself and decline the global competition with her own social model, made up with solidarity, inclusivity, and listening.

Can Naples fear territorial competition? No, we should not be afraid. By comparing cities with fewer troubles, better facilities, and more industries, it seems that Naples should be worried. But it is not like this. It is from the ideas of Dora Francese's book that the answer comes, the benefit to the city that I administrate: regeneration, biocompatibility, and ecosustainability. Let us leave the Asiatic tigers to build their own skyscrapers designed by the *Archistars*. The Petronas towers are not the future for Naples. Our future is not building more and more, but regenerating and valorizing the present. With tender and well-aimed interventions that will provide aspiration, light, and splendor to a precious architectural heritage, which goes from baroque to rationalism.

A new planning season is needed, which will make the city free, by means of pedestrian paths, of new urban spaces, and provided with a sense that is proper and true, being rooted in the bimillenary history of Naples, to escape from those "non-places," by quoting Marc Auge, which too often have filled our cities. Naples should not be filled, but opened. There is no need to build, but to regenerate. It is time to abandon the cannibal architecture that has consumed soil, space, and value, which is physiologically alien to Naples, a city whose first architect was Mother Nature.

Therefore, Dora Francese's book is particularly important. We need, also at a political level, to build a new sustainable paradigm for the urban development, which will reduce the *ex novo* interventions and valorize the existing cultural heritage. Relying on urban regeneration, conscious of the need of renovating, but comprehending the cultural references to which architecture and planning should register. In fact, we feel, more and more, the need for creating a solution system suitable and applicable to the reality of the Mediterranean region's traditional culture, conscious that only by protecting the amenity of our urban context, we can win the challenges of globalization and remain local.

Naples is poor in physical heritage but rich in cultural heritage. Poor in tangible goods but rich in immaterial goods. Our challenge will be won because eating a Neapolitan pizza in a small street, which follows the trace of an ancient Roman road, between a gothic arch and a baroque column, with the street smells and the sound of a mandolin in the background represents a unique value: a value that does not have a price.

Luigi de Magistris
Mayor of Naples

Introduction

The environmental design of a reduced footprint
Bioregionalist technologies and materials

It is actually in the places that the human experience is shaped, is stored and shared, and its sense is processed, assimilated and negotiated. And it is in the places, and thanks to the places, that the desires develop, take shape, feed by the hope of being realized.

(Zygmut Bauman)

While presenting and introducing this book, a brief background of the cultural situation, in which it is framed, is necessary, besides an indication of the goals and content.

The field of knowledge and intellectual interest for the city and for the whole of man-made territory has been studied, tested, and deeply investigated by a number of disciplines and a variety of experts, all aimed at discovering the link between nature and human work.

By canalizing the river course, by manipulating the water for irrigative purposes, by clothing the hills with trees, by restoring health to an infested territory, man is pushed to adapt to places, is compelled to mold it according to his proper social organization.

(Bevilacqua, 1996: 5)

Any new literature, produced with the aim of making the sustainable approach to urban development practical and professional, will promote a more active debate, which stops relying on the concept of unlimited exponential growth. The cities in Europe and the rest of the world could finally start thinking about different development. In order to make design choices in harmony with space and time, it is necessary to confirm Norberg-Schulz's words:

Architecture consists of meanings, [as also the bioregionalism declares] rather than practical functions ... [which] are defined as existential ... [in fact] one of man's fundamental needs is that of testing the meanings of the surrounding milieu. When this is realized, the space becomes a set of places.

(Norberg-Schultz, 1962: 21)

The material culture should then be in harmony with the social and spiritual values of an ethnic community, so applying a revision of the conventional constructive systems, which in the postindustrial age has altered the interface with place, by approaching architecture mainly as an art form, which can be abstract and cut off from economic and human values.

A global warning is required to stabilize the catastrophic increase of greenhouse gases (the increase is, for the example of the carbon dioxide equivalent gases, assessed at 650 parts per million volume). This is pushing the majority of OECD nations to begin to make draconian emission reductions in the next few years (the reduction should be at least of 6 percent per year). Unless the world can reconcile economic growth with decarbonisation, this will require a planned economic recession.

Another solution relies on the concept of *eco-equity*. Europeans use two to three times and North Americans use three to four times their equitable share of world biocapacity and contribute proportionately more to humanity's eco-debt. Global biocapacity has been assessed as approximately 12 billion hectares per year, while the current median human eco-footprint is measured as 19 billion hectares. The overshoot is approximately 58 percent.

All these occurrences lead to a great social imbalance in eco-diversity. Eco-apartheid is a contemporary reality, through which the world's rich competitively excludes the poor. The rich live in the world's healthiest, most productive habitats. Impoverished people and racial minorities are confined to urban slums and degraded landscapes that are characterized by toxic waste, polluted air, and contaminated food and water (see Wackernagel, 2005; Rees, 2011). This includes the physically and socially decayed outskirts, as well as some parts of urban areas.

"Our present life is no longer compatible with the maintenance of ecological cycles, required for living species' survival, man included. The present productive engine dismounts the environment…Leaving the individual enterprise does not bring any advantage, if…the *common good* is compromised" (Tozzi, 2013: 10), as the "economy is a sub-system of the ecology" (Masullo, 2013: 13). The question of economic growth is also linked to the population's growth, but it should be carefully compared with the natural resources' availability.

> The high population densities, the rapid consumption of pro-capite energy and matter and the increasing dependence on commerce…lead to the fact that the ecological localization of human settlements no longer match their geographical placing. For surviving and growing, the modern cities and industrialized regions depend on a wider and wider global area, made up of productive territories.
>
> (Wackernagel and Rees, 1996: 216)

Thus, the new theory founded by Herman Daly, called the *Bioeconomy* (biological economics), has provided some principles, now universally approved and shared. They are the *sustainable efficiency principle* (the renewable resources should be consumed at such a speed as to allow nature to recover them) and the *carrying capacity principle* (the production of the goods should not provide waste, refuse, and pollution that cannot be absorbed by the ecosystem within a short time; there should not be storing effects) (Daly *et al.*, 1997). The only way to satisfy these principles will be through trying again to be in harmony with natural cycles, mainly while transforming the territories and constructing new architecture.

> In the relationship between the construction and the nature, essential was the role of the metaphysic. The architecture floats in between the sign, i.e. the metaphysic being reason, and the means, that is the effectual being reason… the architecture is at the same time cosmic and mechanic.
>
> (Semerari, 2003)

Today, it is important to underline the application of such a sustainable approach that can reduce the new interventions and valorize the existing cultural landscape, not only to architecture but also to urban development. This book aims to contribute to the creation of a consciousness of the role technology plays within the subjects of sustainable urban regeneration. "The techniques had engrossed the poetry borders; it had neither built horizons, neither killed the space, nor imprisoned the poets. In any instant, from the technique progress, dream and poetry come out" (Le Corbusier, 1946).

> If our huge population is to survive in decent living conditions, if we are to achieve true sustainability rather than just lower fossil fuel consumption and emissions, then we must profoundly change how we live and relate upon the earth. But to stop merely camping or picnicking upon the planet as irresponsible transients – to truly dwell upon, belong to and feel reverential connection with the earth – we once again need to make environments conducive to this: real places. One way we will know we are getting there is when there are no longer the meaningless and unloved residual spaces that surround modern and contemporary buildings, and instead even the smallest spaces are proper places, and cherished for that.
>
> (Buchanan, 2012)

The place becomes a new matter for constructing, and the design is the action of throwing forward a new energy. No other system of constructing and transforming the world will again be possible unless we start and continue to think at our local realm as a whole. The solution comes from a number of existing theories and movements, which promote the local and territorial financing as well as resources, that is, the bioregionalism and the de-growth. Within the book, both the definition of technological solutions for open spaces in the city and the chosen practical examples will reflect these approaches.

The debate about a zero kilometer architecture, construction, and project restrains the field to the question of fossil fuel consumption, necessary for the transportation of systems and products. Instead, it is necessary to activate a wider concept of sustainability that will include a holistic way for many sectors of the building market. Today, the following applies:

> Modernity has reached the point of being tackled with its proper limits … the modern life seems to be the only possibility … that makes it difficult for humanity to adapt and to assume choices which could be coherent with nature.
>
> (Bauman, 2013: 179)

> The modernity is the combination of three elements: the faith in the potential of science and technology to be able to solve all our problems; the belief in the State as combination of power and policy, and the global capitalistic economy, with both goods and information exchange. … The old habits, customs and routines which we used to apply with successful effect, no longer work, while the new life-styles, becoming more and more efficient, are still in the experimental stage of design, at the drawing board, and it is not yet clear which ones are reliable and selectable…. We are in this state when modernity faces its limits.
>
> (Bauman, 2013: 180–1)

The solution is dialogue, not as argument but as interaction and solution provision, leading to cooperation, where there will be neither winners nor losers. The open spaces of the urban context are a good field of application for these new trends, for they represent a great hazard to our habitat.

> Beyond the cases of excellence, where environmental, structural, besides symbolic, qualities harmonically converge, it is difficult to track down, in the contemporary constructing praxis, such an equilibrium between the parts and a following succession of urban open spaces, mainly in terms of fruition.
>
> (Carbone, 2009: 72)

> The question of open public space … is one of these new hybrids that helps overcome old conflicts of representation. Whether we call it blue and green ways, natural infrastructure or open space, it can play the same role as the streets and squares of the *real city* towards positive sub-urbanity.
>
> (Vanier, 2011)

"A neutral ground (the public space) does not abstractly exist, but agreement and conflict grounds, from which the sense of belonging is born, exist" (Persico, 2013: 13). Therefore, it will be possible

> to hypothesize that the difficulties of land could be overcome if the concept of the *City of Projects* will be changed into the concept of *Regenerated City*

which supposes the identification of a new social root as palimpsest for a territorial texture able to produce both economic and social values.

(Persico, 2013: 12)

The regeneration of open spaces requires a number of actions, such as *requali-fication*, that is, to return quality to the settlement and areas, by upgrading performance levels, the adequacy of the functional requirement and proximity, and
avoiding obsolescence. Also, to provide *revitalization* by indicating and creating
new use destinations that will attract people, tourists, and citizens; thus, returning
life to the places, and finally, *to valorize* the benefits, identity, and the landscape
chances and potential as much as possible, by making a soft project and avoiding
aggressive ideas and technologies that can disguise or cancel the identity.

This book is not meant to declare the subtle boundary between the architecture
and planning disciplines, it is not a means either to describe the contents or the
practice of the urban design expertise; it is not even aimed at building a theory or
a general idea of a sustainable urban regeneration *tout-court*. The book consists
neither in a landscape theory nor in urban programming and planning proposals.
The main *objective* is, instead, to explain and to share some practical and actual
procedures that will suggest the appropriate use of definite technologies, specific
materials, and updated products and byproducts in the new design practice for the
regeneration of some parts of towns.

The urban and peri-urban areas are used as case studies to demonstrate that,
although the political, strategic, and planning programs for the city could be highly
responsible, as far as sustainable, social, and ecological items are concerned, the
people's satisfaction and city values could be achieved only if both the technological solutions and the environmental design approaches are selected according to
a deep understanding of the existing conditions.

Taking account of the climatic milieu, vegetation, quality of air, and water and
soil, the design for regenerating the urban as well as peri-urban areas would actually contribute to providing a higher level of welfare, comfort, and quality of life.
Thus, promoting both community income and valorization of the built heritage.

This book is meant to emphasize the technological and material, rather than
merely shape and planning, considerations of the regeneration processes. One of
the aims is underlining the importance (and role) *bioregionalism* plays in technological choices, strategic decisions, and design procedures within environmental
transformations.

Another aim of this book is to focus on those solutions that would be appropriate to the cultural tradition of the Mediterranean region, trying to avoid the use
of northern European strategies and materials, which are not adapted to southern
realities. Searching for a system of solutions appropriate and applicable to the
reality of the traditional culture of the Mediterranean region requires the study
and definition of levels of ecosustainability and biocompatibility, as well as of
participation and performances.

This book is trying to prevent architecture from being egocentric and performing
as if it was a means for the designer to become rich and famous. It is a reminder

to all of us of the main task of the architects, engineers, industrial designers, and planners, that is: reducing the ecological footprint and satisfying the requirements and desires of the earth's inhabitants. It is necessary to make "a step backward as regards the attention-seeking of the one who builds" (Peregalli, 2010), as "throughout the 1970s there was an exacerbation of stylistic concerns at the expense of programmatic ones and a reduction of architecture as form of knowledge to architecture as knowledge of form" (Tschumi, 1994: 140).

The content of the book makes an effort to provide a few examples of practical employment of the sustainable technologies in the future scenario for architecture within the urban and peri-urban realms. The book is divided into three sections, in the first of which, the cultural background, present trends, and some new conceptual methodological approaches to the questions are outlined, according to both the culture of the technological design procedure and the *bioregionalist* principles.

Following the first part dedicated to exploring the updated and innovative viewpoints for the regeneration of urban open spaces, the second and third parts will describe some contextualized samples of urban regeneration. They were chosen among a number of design practices, either built, proposed, or studied by the architectural students of recent years, in which sustainable technologies are utilized in the urban and peri-urban realms. The examples shown represent an application of the methodological proposals and ideas as valorization and conservation tools, and as incentive for the citizens towards a better livability, which at the same time, can allow them to stop thinking and behaving in a consumerist way.

Part I

Investigation of the urban systems according to the technological approach to the environmental design

1 A new strategy for the city

Sustainable regeneration

a The sustainable city

The ancient definition of a city, etymologically derived from the Latin *civitas*, suggests that the urban settlement should mainly be configured as an organization of human community, where the habitat, as well as the people, could retain the same importance and the same right of being part of the settlement itself. According to this idea, the city is a *mixité* of a number of structures, such as the land, the citizens, the activities, the infrastructures, the roads, the squares, and the parks. Even in literature, the idea of city is meant as a combination of human activities and physical infrastructures:

> We do not...provide specific importance to the name of a city. All the metropolitan areas were built by irregularities, turn overs, drops, intermittences, collusions of things and events, and, in between, bottomless silence spots, by railways and virgin soils, by a great rhythmical beat and by the eternal disaccord and disarray of any rhythm; and as a whole it was similar to a re-boiling bladder set in a material container of houses, laws, regulations and historic traditions.
>
> (Musil, 1957: 6)

Therefore, one can describe the city as a human ecosystem, part of the wider built landscape, and as such, can be "sectioned in the bio- and socio-diversity of the context." If, instead, we consider the city and its built landscape as a mechanism, the landscape is "produced starting from the elements borrowed by other contexts" (Raffestein, 2003).

This model of the compact city can instead be described by numbering the various components, as outlined by Evans and Foord:

> The notion of compact city...encompasses: 1. Social mix (income, housing tenure, demography, visitors, lifestyles); 2. Economic mix (activity, industry, scales from micro to large, consumption and production); 3. Physical land-use mix (planning use class, vertical and horizontal, amenity/open space); 4. Temporal mix of items from 1 to 3 (24 hour economy, shared use of

premises/space – e.g. street markets, entertainment, live work). These urban environmental elements, that combine to determine the quality of life in higher-density, mixed use locations, can be triangulated in terms of the key features.

(Evans and Foord, 2007: 95–6)

The key features mentioned are physical, social and economic. Provided that no city can exist without its human heritage, the importance of its territorial boundaries and tangible built areas cannot be forgotten.

This chapter is meant to underline the importance of generating a new opportunity for the city. Rather than discussing the planning methodologies, or shaping design for the modern and old city, the imagination of various technological solutions can show, to both citizens and guests, how to interact with the existing structures and how to preview a different use of the city spaces. At the same time, it can create the chance of proposing various models for eventual new processes. These new processes cannot necessarily be identified with new constructions, new buildings, or new assets of the city. On the contrary, the idea supports the existing facilities with new use and new arrangements: only in this way, the real sustainable policies can be achieved. In fact, to avoid employing new materials, new energy provision and waste production can be considered as the only ecological means of facing the city question: this is the point of view of the environmental design response to the city strategy resolutions.

As far as the sustainable city is concerned, what is this chapter's point of view? The most important parameters that a sustainable city should respect can be listed easily, while the parameters of how they could actually be applied and could be maintained in time are more complicated to understand. The list includes:

- rational use of resources
- employment of renewable energy (solar, wind, geothermal, etc.) according to the environmental context
- use of passive and active sustainable technologies according to the climatic context; attention to the "genius loci" (natural landscape and traditional architecture)
- earth care (low carbon emissions, low pollution, no greenhouse gasses, no ozone depleting emissions)
- attention to the users' needs
- circular approach (using lifecycle assessment ex-ante methods and promoting the recycling and reusing of any activity products)
- working with the climate (exposure to sun and wind of activities)
- provision of green areas
- nearness to water supply (for drinking and other uses)
- connection to nonurban areas
- human health care (use of non-toxic and harmless materials)
- waste treatment (see Figure 1.1).

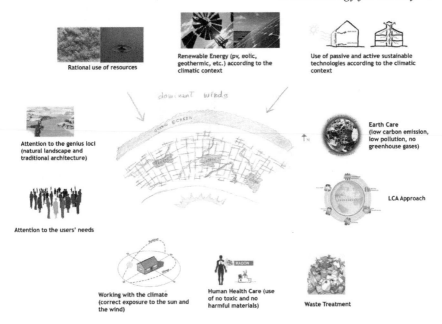

Figure 1.1 The sustainable city

The modes to achieve the completion of these parameters can be synthetized in three main actions: the decrease of resources, the minor possible constructions, and the reduction of waste. This means paying more attention to energy, material, and water, while proposing possible strategies and policies for any eventual transformation. The selection of design procedures should thus be in line with these principles and can be easily achieved through three main steps of city life: building, using, and rejecting. In each of these activities, the sustainability can be measured according to the amount of resources included. Instead of manufacturing a new concrete object, such as a building, a road, a square, or simply a bench for a park, the sustainable solutions for an ecological city could be the following:

- converting existing things
- modifying the energy consumption of the system
- reducing the nonvirtuous cycles
- reducing the amount of matter, energy and water needed for the transformation
- previewing a type of use for the new solution that would require low energy, less raw prime matter, less use of water, and less engagement
- attempting to reduce the unrecyclable or unrecoverable waste.

As it is known this is not the first time that such a policy has been advanced. Currently, for many institutions (the international agencies, the European Community, the local associations and committees) dedicated to saving the planet, the question of employment of resources is key, because even the survival of protected animal and vegetable species depends on that, and, of course, the preservation of human community.

Nonetheless, the tools that are useful for achieving these results have not become a real heritage for any professional or local authority making use of ecological and biocompatible instruments, both as methodologies and technologies. Scholars and land administrators have developed various approaches to resolve the city transformation question, with regard to the eco-cities, the sample-city, the mixed urban use, and the synergy-city.

They start from the fundamental concept that a different way of inhabiting this earth should be imprinted on the means of dialogue, shared choices and respect for users' as well as the needs of the environment, and never on aggression, war, and prevarication neither to the individual nor to nature (see Ravetz, 2014).

This idea of de-growth,[1] is not new: even the Chinese philosopher, Lao-tzu, in the sixth century BC, declared that we do not need big growth, but a great vision and simplicity. The more updated urban studies tackle the needs of de-growth, as the primary aim is not only the order and planning of the territory, but also of the *urbe,* which represents the blending of land and joining of citizens. As Bernardo Secchi (2014) states:

> Usually, the debate on the de-growth ... is right for the economists, as far as the increase over time of the rate is concerned, it is right for the sociologists, when its unbending or stopping explanations are interested...The planners should only acknowledge it, consider it, and consequently modify their design procedures, their idea of land and city.

This indicates that the city should be modified, re-arranged, regenerated, or even re-created without changing the amount of used material and energy. He is also optimistic regarding the possibility of applying this change into the modern city, when he hopes that "something is moving in this direction...due, on one hand, to the number of researches" that are currently based on these subjects, and on the other hand, to the fact that

> more and more frequently designers are asked to study and process some *visions*. Visions rather than regulations, recognitions of meta-preferences, which...emerge from the whole of the researches ... non final pictures, but paths, trajectories which explore the possible...new fictions which can provide a perspective of the contemporary society.
>
> (Secchi, 2014: 313)

These visions can then be identified with simple solutions that will employ existing objects and parts of the city. At the same time, they could promote a strong and radical change in the life of people because the absence of construction and the presence of creativity and projectuality have properly substituted the aggressive speculations that had characterized the last century.

The sustainable city is one that does not promote any increase, any growth or any additional constructive activity, but rather only carries on its development, improvement and regeneration.

b The city regeneration question

Regeneration can actually be reconsidered as a method for reproviding comfort and livability to people inhabiting a town and to tourists as well as neighbors visiting these places. To "re-generate" means to give birth again to something, to provide to the existing reality a renewed life, new genes, new importance, and alliance with the earth.

Within this updated concept, there is the idea that this new life does not need to be created and delivered by employing a great amount of matter, energy, or water. No new resource is necessary, on the contrary, the existing sites, cities, and way of living could improve their vitality and appeal, without increasing the environmental load, both locally and globally. As Aimaro Isola declared when asked about the alliance with the earth of a design process or a landscape construction:

> [It is] a very difficult alliance that with the earth…while designing, a colloquium begins and develops during which, slowly-slowly, we notice that the place reflects our face, we assume its faults and its virtues…Within the constraint of the built environment… [a number of] contradictions are hidden and co-living. We all would like…that in our works the violence would be sedate; so we could, by reflecting ourselves into the places, see a new happy face of ours.
>
> (Isola, in Priori, 2012)

It became clear during this interview that it is the place itself that carries the potential of its transformation, improvement, and increase of quality. The regeneration of a place, which would be in alliance with the earth, should, in any case, not be violent or aggressive, but respectful of the natural existing elements.

How can this be possible, without low and soft interventions? Possibly with sustainable and respectful technologies or only a light reorganization and reuse of existing objects and townscapes?

A crucial aspect of the modern technological design is the difference between natural and artificial objects, which cannot actually be simply described in the twenty-first century as everything that is human-made, including the animals and the plants. Aimaro states:

> Nature stays in front of us, not as a different other being, but supported by ineludibly rules, mathematical, geometrical, growing, of dissolution, of dissipation…and thus nothing remains to us than being wrecked,[2] softly, poetically, Leopardian? Or, on the contrary, do we actually make part of the nature, is it us and not only our body, but also our thought, our home, the landscape, the world? Therefore everything is nature, thus nothing is nature?

> Nature according to Spinoza: *naturans, naturata*?…where does nature end and where do we start? Or better, how do we succeed to conquer within a space of freedom? Space which could provide only the poem? But which could really open with the project a space to life?
>
> (Isola, in Priori, 2012)

In these very suggestive sentences, Isola creates the space for design within the great realm of nature, without destroying the peculiar characteristics of nature itself. However, at the same time providing a number of important reasons for helping the natural landscape to host humans with their requirements, their life arrangements, personal qualities, and their "poetry." An important item to be safe-guarded by two actors of the architectural and technological process, the designers and the users. Both should try an alliance with nature, and both should neither deny nor neglect their poetry, to be closer and not further away from nature itself.

It is in the very city that the citizens and the designers meet and produce their vitality, their poems, and their suggestions. The softer the touch, the less detach-ment from nature there will be, both from the biotic and nonbiotic.

The city is made up with buildings, big squares, big parks, large streets…but all these important elements are also separated by small roads, little mews, and narrow paths. Narrow paths should be considered as an important part of the European and Mediterranean city, whereas larger streets are more common in the USA.

Therefore, regeneration should also become an opportunity for reflecting upon these minor realities rather than underlining the importance and relevance of the monumental and gorgeous pieces of the city.

c Methods of regeneration

The means adopted by this approach to regeneration are contained in a great number of literature essays, which assert the importance of reducing concrete interventions, while increasing reuse, recycling, and recovery of existing elements, from build-ings to small trees.

In the last few years, various methods for regeneration have been processed and diffused. These methods pay "careful attention to the quality of the street-scape and to enhancing the network of small spaces and squares implicit in the historical layout" of the ancient cities, thus "…creating an attractive and popular place to live, work and socialize" (Thwaites *et al.*, 2007).

A socially and sustainably correct method for sensitive architectural regenera-tion process of an old city would provide a number of positive characteristics and features, such as a "relationship with other city districts, exploitation of eventual waterside areas, vibrant social opportunities" (Thwaites *et al.*, 2007: 5).

The great goal of the twenty-first century is that of environmental preservation. The need to provide ecological quality of life. Therefore, "in order to create com-munities, besides the social and political commitments, it will be indispensable *to provide climatic, acoustic, non-polluting and visual comfort to people*" (Thwaites *et al.*, 2007: 5). Thus, regeneration takes the meaning of renaturalization, even if this concept cannot be applied wholly as it is used for plants and agronomic issues. Being in an anthropic milieu, renaturalization regives identity and history to the site and provides to the people a sense of home and recognition. This can be achieved by respecting their requirements, both expressed and unexpressed. "A successful neighbourhood should also meet long-term needs – the cycle of a lifetime – in addition to daily needs" (Thwaites *et al.*, 2007: 5). During necessary transformation, aimed at achieving the aforesaid users' requirements in the built landscape—already existing and preoccupying the urban and peri-urban place—the methodology has to take into account the fact that "the built landscape mobilizes…the three worlds of the matter, of the science and of the senses: all the built landscape is the converging spot of these three worlds" (Raffestein, 2003: 31).

The city, as part of territory, requires the same modalities of regeneration. The territoriality is defined as:

> The whole of relationships which a society – and thus the individuals which belong to it – interchanges with the exteriority and with the other from itself, in order to satisfy its needs with the aid of mediators in the perspective of acquiring the most possible great authority, in providing the system's resources.
>
> (Raffestein, 2003: 31)

Therefore, the city should be requalified according to this great number of relationships, both physical and anthropic. In fact, the people, as well as the space, can be considered as a visual and sensitive part of the requalification process, as far as socially sustainable urban design is concerned. The regeneration work is a phenomenon that nowadays invests wide parts of European cities.

> A cursory look at…any city skyline today reveals a plethora of cranes signalling a rush of construction which aims to regenerate neglected city quarters, often removing or re-inventing now obsolete industrial usage and creating smart residential, retail, leisure and commercial attraction to re-populate urban areas.
>
> (Thwaites *et al.*, 2007: 4)

In many European countries the "city skylines are changing, dereliction and neglect are being transformed, people are returning to cities to work, play and live" (Thwaites *et al.*, 2007: 5), whereas in Italy, and more generally in a number of Mediterranean countries, the need for regeneration is still strong in many towns, and people are eager to receive the deserved quality of life.

One of the possible modes for a regeneration process is the Roger's task force, that is, "to breathe new life into towns and cities generating a strong sense of place" (Thwaites *et al.*, 2007: 5). The concept of place, which is very different from the meaning of any site or location, is actually part of the city, and thus, requires a complex process of regeneration, carried out by means of master plans that are produced in order to emphasize "spatial rather than merely planning considerations" (Thwaites *et al.*, 2007: 5).

The methodology for achieving a sustainable regeneration today requires the need to think about a number of architectural solutions "...able to allow the reproducibility of resources and the regeneration of spaces, as it happens in nature in the vegetable world" (Allen, 2008). This process could be provided only when professional tasks and administrative procedures will show consideration to the environment as well as to citizens, and attempt to reintegrate urban phenomena within the natural cycles, thus recreating new life and regeneration.

d The role of architecture and design

"The architecture is inextricably connected with the idea of a city for the citizens" (Tschumi, 1994); thus the place, where the activities as well as the citizens' life itself are running, comes obviously from the empty spaces, internal and external, created by edified fabrics. If at the same time "no architecture can exist without an historic, geographic as well as cultural context" (Tschumi, 1994), the dialectic and interchangeable relationship between architecture and environment will result as the only one able to provide a cultural and social justification to the idea of city.

During the studies for city regeneration, it is not only the planning and urban legislation logic that can contribute to clarify the goals and the successful actions, but also the studies aimed at understanding the procedural paths, which had led to the physical and constructive arrangement of the very urban area.

As it is well known, the city, as agglomerated of various realities, can promote sustainable routes during the management and maintenance processes, or alternatively, can be oriented towards strategies that are less respectful of natural and cultural ecosystems.

Architecture, as a science that studies and promotes the art of building, could follow strategies and methodologies aimed at the protection of the environment, or alternatively, follow less complex paths, less coercive and more linked to the designer's imagination and freedom. In this way, any proposal for a built and anthropic area is legitimated, as long as it met the minimum requirements of legislations and livability of final users. If the first route is followed, and thus the strategies match the ecosustainable and biocompatible criteria, then the derived architecture could assume a good environmental quality.

The low quality of architecture determines reductions in the social realm, as the physical decay includes a complication in adopting cooperating systems, social exchanges and additional services provision. The key question concerns how the architecture could work on the buildings and the urban objects without considering the social items of a city that aims to achieve sustainability, solidarity, and global issues. One method has been identified by a number of designers, who question the social issues versus the economic and financial subjects on a daily basis. Whilst guaranteeing social benefits and quality to architecture, and achieving users' requirements (such as comfort, fruition, security, safety, and so on), a number of features and facilities are needed, besides a certain amount of investment, which may not always be paid back in monetary purposes, but very often in terms of quality of life, of durability, of energy saving, and of health.

Conversely, the designers, especially when they deal with public administration, for example during social housing provision, have to deal with scarcity of financial resources. In this scenario, not only is the sustainability, as a whole, difficult to achieve and to be globally pursued, but there is the risk that even the more elementary needs will not be met.

How can it be possible in these circumstances to facilitate social sustainability and participation processes? One of the very well-known architects, who has always been caring and fighting for human rights and social needs, is Alejandro Aravena from South America. In order to find a partial solution to the previously mentioned problem, he responds about his design team.

> [The] starting point is always to get to the nitty-gritty of the problem. How much money is available? What are the social, political, economic and building conditions? How is the market? The data are wanted according to what they are. The project entails no personal creative theories bound up with a vision of space or self-expression. Theory pops up the moment we don't need greater insight.
>
> (Aravena, 2009: 113)

Thus, the question can be solved with a different methodology, which takes into account from the very beginning, both the environmental and the anthropic values and desires. In addition, there are the circumstances—economic, social, and physical—by means of the deep and conscious knowledge, which will provide the right decisional procedure for building new architecture, for requalifying the existing, and for integrating into urban sites. Thus, the ideals, theory, and single designer's egocentric process should be separated and reality will emerge and create the answer: "but it is the theory tied up to the [reality] which serves to optimize our response to them, not to [promote] ideal solutions" Aravena (2009: 114).

This very question of the antagonism/alliance between economic and social items, central in any sustainable process, becomes even more dangerous when operating within the city realm. The architecture's role will be interacting between public users, customers, and regional administrators, in order to interpret local needs (social, economic, and environmental) to create a context of synergy and cosharing of responsibilities.

> The architect is valorised as far as his power of synthesis and visualizer of urban locations are concerned, and on the other hand, he is required by various levels and fields of power so as to state spatially such values often in contrast with each other's…The risk is that of reducing the architectural comprehension to a simple servant of advertisement, a copywriter of urban policies already established, and that of exploiting his art craft so as to solve time by time some spatial argument of a little importance.
>
> (delli Ponti, 2012)

The role of architecture can be considered as responsible for city transformation, either for positive or for negative effects: "if architecture forgets the big questions of the environment – the energy consumption, the rationalization of processes – it will surely contribute to the unbalance, thus [the architecture itself can be considered as] one of the key factors of the environment" (Portoghesi, in Seminario, 2009: 43). Portoghesi insists on the fact that the responsibility is not only due to ignorance of sustainable goals, but to the application of them into design procedures.

> The trouble does not reside in the fact that the architectural culture ignores the environment's issues, but in the fact that it tries to solve them not as a whole, and without introducing a sensitivity…The question is not only if to use or not the photovoltaic panels, but it is to express, through the architecture, this new sensitivity towards the values and the hazards for the environment.
>
> (Portoghesi, in Seminario, 2009: 43)

Some of the architects and architecture historians do believe that the role of architecture could be less coercive and less strong, while dealing with environmental issues to avoid the construction and transformation aggression. However, without the architecture *tout-court*, not even ideas, concept, spatial solutions, and place recognition and identity care could be achieved. If, in fact, "it seems that architectural culture gets in the way of architecture really fulfilling its role in the environment," conversely, a possible way of avoiding the egocentrism of designers is possible. "Wines' explicit programme of de-architecturing architecture" will be a way of reclaiming the right role of architecture to the city asset, rather than only to its image. Trying to manage with a new vision of architecture means that it should be avoided to produce any "shapist architecture – the kind of thing that has flourished in architecture deconstruction onward – [which] is simply re-running an outmoded formal idea of sculpture" (Jacob, 2009).

This does not mean that architecture should abandon its idea and specific character of being an artistic activity. On the contrary, it should make a big difference between art and work to know when it is possible to have ingenious and spatial ideas to propose and when it is necessary to employ simply an existent product.

> For Lefebvre [in fact], a work is a work of art, whereas a product is the reproducible result of a mechanical process. Venice … is a work. Today, urban space is a product of industry and politics. For example, the mass-producing industrial economy has translated the dream of a single family home in the only way it could, into large-scale suburban sprawl.
>
> (Lefebvre, 1991)

Therefore, architecture is necessary more than ever to avoid the city becoming a machine and a sum of products, and so pushing even more towards the consumerist society. In fact, "zoning and related development controls which enable uniform suburban subdivisions are political abstractions that are used by politics to subjugate space…Abstraction serves to alienate people from space and to facilitate the exercise of political power over it" (Lefebvre, 1991).

The architecture's role within the city regeneration procedures, as will be explained in more detail, will be complex and multi-faceted. It will regard the social as well as the policy questions, it will be imperative for environmental sustainability, it will be necessary for guaranteeing people's welfare, livability, space recognition, and finally, promoting less construction and more art into our lives. Less construction means avoiding being aggressive towards nature.

e The role of the sustainable technologies

Within the methodologies and the procedures aimed at urban sustainable regeneration, the fundamental role of the technologies is considered as a measure for applying concrete and tangible solutions to the designers' ideas, which over time are performed in the desired direction of a minor impact and a minor ecological footprint. The constructive and the virtual technical procedures, as well as the materials, the energy shapes, and the hydric use, in fact, represent the sole response to any possible urban regeneration. The sustainable technologies can be defined according to their capacity of composing architectural systems at high ecological content and respecting the urban ecosystems, both in their natural and cultural elements.

Techniques have often been considered as a negative objective rather than a means, and their influence and necessary contribution have been taken as restrictions to architects' freedom, while at other times, the designers have depended on the potential of the same technical innovations.

> Throughout the present century architects have made fetishes of technological and scientific concepts out of context and have been disappointed by them when they grew according to the process of technical development, not according to the hopes of architects.
>
> (Banham, 1960)

Another important role of technological choices, within the process for sustainably regenerating the city, is that of including the chance of transformation (technology is the transformation means according to Ciribini's (1984) words) as a possible easy, soft, and respectful solution, not necessarily hard, aggressive, and mechanical. Materials, products and systems can be re-aggregated without involving new construction, and this procedure requires a great amount of technological choices, mainly focused on the absence of polluting effect, the respect for identity of the place, the reduction of energy requirements, and so on.

If the technology is meant in this way, then any innovation in technology will be useful in order to provide more livability and more sustainability for a city's users. Even though some "generation ago, it was the *Machine* that let architects down, tomorrow or the day after it will be the *Computer* or *Cybernetics* or *Topology*" (Banham, 1960), and architects are always fighting against the contribution of the techniques. In fact, some believe that "what [they] have hitherto understood as architecture, and what [they] are beginning to understand of technology are incompatible disciplines" (McHale, 1967).

The vision of technology's role within the design process has often been debated among architects, engineers, and planners. They have often believed that *technological* meant *modern* and nontechnological was something more linked to tradition. According to technological studies carried out since the 1960s in Europe and in the USA, the technological means can be classified exactly according to their content of aggressiveness towards nature, and thus, divided into soft, passive, green, and appropriate, or into hard, aggressive, alienating, and unsuitable.

The technical instruments that provide a design with the tools for actuating the proposed and thought transformations are to be again reviewed. As we know the traditional technologies were often very low polluting and appropriate to the site. This leads to the concept that tradition had its proper technologies and that modernity can still occupy that field of technologies that provide limited damage to our planet. Banham, an ancestor of the soft technologies, and who has alerted us over pollution and aggression to the natural environment by the construction sector, still considers the architectural profession "torn between *tradition* and *technology*" or as he phrased it *"history and science*...needed to re-define its limits in the midst of these competing bids for intellectual domination." By *tradition*, Banham means the stock of *"general professional knowledge*; by *technology*, its opposite, *the exploration of potential through science*" (Vidler, 2012).

In this way, the two terms acquire a very peculiar significance and try to report the technology to its proper sense of a bridge between scientific discovery and concrete architectural work.

The specific role of the sustainable technology, within the design of city regeneration, is controlling the achievement of the environmental issues, by means of selection of those procedures, strategies, and materials, which best suit the project and the proposal.

The idea is introduced here that these technologies should be as less artificial and aggressive as possible. They should become more a way of transforming and converting the existing urban objects, architecture products, and facilities in the public open spaces rather than a system for constructing new elements and buildings.

If the technology is defined as a transformation means, then its role can be expressed within three different systems: during the process of planning-design-construction-disposal, during the application of materials and products, thus considering the necessary steps and instruments, and finally, as a tool for changing the prime matter into products or existing constructing elements into a different system.

Thus, sustainability can be found in the various steps of the process and procedures and in the selection of tools, products, and materials.

f The needed scientific investigation

Any possible solution towards a sustainable city starts with the need to acknowledge the present situation. Either the aim of the study is to improve and increase the performance level of the existing establishment or it is to operate big or small changes to renovate and transform both the social and the built environment. It is necessary to clarify the present situation by processing an investigation frame aimed at facilitating the proposal of more appropriate new shapes and configurations. The required information mainly considers the environmental, social, political, and economic resources present in the site under study. The natural resources still present in the analyzed urban space should be taken into account. It is well known that, even when the city is completely artificial and not even a tree can be found, the climate, the air, the sound milieu, and the inhabitants are nonetheless to be considered as *natural*. An important issue within the requalification process is that of knowledge: "the observation levels of the built landscape [are] to be assessed in order to prevent [that] any dangerous intervention should be [applied]" (Raffestein, 2003).

When interpreting and reading urban areas according to environmental design procedure, it is essential to take into account the comfort conditions. Consequently, a number of analyses are needed on both the original situation and the present changes, as far as materials' and products' performances are concerned.

The direct or indirect observation of the phenomena occurring in these urban areas allows us to know their effects on users' health and on the environment's wellbeing. By interfacing the environmental behavior of urban areas with design decisions, several scientific investigations are needed, which should be calibrated according to complex system characters.

The preliminary action aimed at assessing comfort performances is to verify the punctual knowledge of technical aspects and of physical and mechanical properties of original materials. By reading both the environmental and technological systems, the potential of exploiting renewable energy sources can be investigated.

During the comparison between the various environmental factors of the site (climate, morphology and *genius loci*, acoustic conditions, air quality, historical as well as physical context, etc.) and the constructive choices of the urban area (the buildings' texture and materials, technological elements, streets and squares' characteristics, etc.), the bioclimatic and environmental behavior of the site is identified and its potential for saving energy, as well as its performance in terms of provision to users of life quality, is achieved.

A complex interpretation system should be established, with a number of criteria founded on two main parameters, the scale (territorial, city, blocks, and surroundings, building texture, single fabrics, details) and the investigation typology joined with the suitable tool system (visual observation of phenomena, questionnaires and interviews, geometrical evaluations, tool measurements, calculative and software simulations, graphic processing, multi-criteria assessments, etc.).

Identifying a model of behavior for the analyzed urban area, which stands for the site's attitudes towards the environmental answer, could be configured as an aided tool during the decisional process of designing solutions. This will be in harmony with the area to be regenerated, with the materials and technologies already existing in the area and with the environmental opportunity of open spaces.

The behavioral model highlights a number of design potentials, but makes restrictions as far as the identity of the existing cultural heritage is concerned. The solutions should not only be congruent with the aspects, image, blending and the usability of the cultural heritage in its entirety, but also with the comfort and quality of life to be provided for the users of open spaces.

By connecting the information obtained by the various typologies of investigation at the different levels, a complex system of relationships is created between the number of playing strengths (environmental phenomena, technological performances, environmental performances, coherence between the various portions of open spaces, etc.), which will be adapted to build a behavioral model at all times appropriate to the specific site's identity.

The number of stages defines the methodology employed for the scientific investigation procedure. First, the identification of the comprehension levels is needed for the performance evaluation affected by the phenomenology that occurs in the major scales. The second stage defines the selection of those procedures that will prove more suitable, reliable, and flexible, as well as less destructive for the site characters and identity. Then, in the third and fourth stages, the processing of the previous studies would be established through a comparison between the different evaluated data provided by the investigations and processing. This in order to proceed to the construction of an assessment sheet for the environmental performances, which in the last stage, will contribute to formulate the specific behavioral model.

If, as already said, a design for a regeneration process starts with a detailed and correct investigation on the existing site, the model could be outlined by a study applied at various levels, from the environmental characteristics of the bioregion in which the urban center is included to the definition of the peculiar properties of the employed materials.

By processing the outputs of the previous investigations on the environmental response of the site, an informative frame of the specific site actualization can be done. This knowledge allows us to identify a possible strategy for the urban area's regeneration, which will respect the historical significance, the environmental value and the building-texture's material culture. "The ecological capacity of the sites is usually ignored and so is the existing ecological network, as if this were not linked to the wide area of reference" (Persico, 2013).

The environmental studies analyze the sites' behavior in terms of those aspects (heat, humidity, sound, air quality, vegetation, ventilation, sun and shading, and so on) that contribute to modifying the comfort and life quality of open spaces, and, differently from an inner space, cannot be provided easily with artificial air conditioning systems.

Even though environmental comfort is identified as a complex condition, which includes a number of physiological, psychological and social aspects, it strictly depends on the material and technical characters of the built milieu. In addition, it depends on the urbanization of the whole area, the typology of the site, the solar and wind orientation, and on a great number of factors, both due to the context and to the indoor qualities.

Both the macroclimatic and anthropic phenomena could be derived from the interpretation of environmental and territorial elements. The geology of the bioregion, as well as the morphological and climatic aspects, provides the interface with the land, landscape, and the anthropic environment. The solar and wind paths and dynamic processes, as well as those derived from the green areas, the acoustic milieu, the visual, and the air quality, can be connected with the site to identify the phenomena occurring at the various scales.

While analyzing an urban area included in a historical center, the study of the building fabrics' texture is essential because it creates the specific microclimatic conditions, through the interaction of the single buildings' environmental behavior. If the blocks are strictly amalgamated, they present a peculiar bioclimatic character, while—when spread—the environmental phenomena happen to be very different. Thus, investigation on the buildings' typology, on the social relationships, on vegetation processes, and on landscape characteristics, as well as on the logistic aspects such as path networking and mobility, can help to identify the technical and environmental performances of the area.

The underlined analysis of urban areas, in fact, is aimed at investigating both environmental background and the current condition of the place. The first stage, the visual study, is carried out by means of visits on site, surveys about the original state in the literature, and interviews with the neighbors.

As a result of the Master Plan regulations and of the National Trust control, a number of restrictions in force should be examined with regard to the location and the future usage of the building. More detailed and scientific analysis concerns three different levels of survey. First, the technological review of the built environment and the relative construction elements and materials employed. Second, a definition of performances of the original system according to the new usage requirements and to the different needs (in particular the comfort needs are to be taken as principal). The third level of the survey takes into account the environmental behavior of the entire urban area that has to be regenerated, starting from an analysis of the factors in the climate, which depends on both the site conditions and on the thermal performances of the neighboring fabrics.

Only after a detailed, complete, scientific, and reliable frame of analysis can the site be acknowledged, and thus, any possible design solution can be processed in order to be suitable to the existing situation.

Notes

1　One of these roads is the *new economic theory*: "The de-growth is…a policy design… that of constructing, in the North as well as in the South, convivial autonomous and sober societies" (Latouche, 2010).
2　"*Naufragar nell'onda*" from the Leopardi poem.

2 The bioregionalist vision of environmental design

a Bioregionalism and sustainability within the building production

The *bioregionalist* approach can contribute to the creation of harmony between the construction sector and other activities within the complex scenario of contemporary society, according to ecosustainability and the EU 2020 package. The latter, a decennial strategy for European Union development, does not only aim at coming out from the crisis, but demands to fill the gap in the present growth model and to create the conditions for a different type of economic development, more intelligent, sustainable and with solidarity. The five goals of this strategy are employment, education, research and innovation, social integration and poverty reduction, and climate and energy. Seven main initiatives are the framework within which the single countries' governments should achieve these goals: innovation, digital economy, employment, youth, industrial policy, poverty, and efficient use of resources.

In order to propose a sustainable development, a large role is played by the construction sector that consumes a great amount of energy. The running costs for new erections and the maintenance and refurbishment of buildings are responsible for a good part of this consumption, while the management costs are also remarkable.

Thus ecosustainability, declined for the completion and use of building systems, includes not only to adopt low-environmental-impact materials and techniques, but also the necessary commitment for the components to combine various productive strategies. Cooperation between local policies and renewable-resources' employment can contribute to the reduction of the ecological footprint, while using new technological systems out of traditional materials so as to promote the relaunch of local production even in the out-of-date sectors.

According to the European Community's goals and to the new strategies for a more sustainable and more livable future, a number of experiments are to be implemented. This will allow the connection between research (both at the university and at other institutional levels) and the building market. In fact, one chance for strengthening the productive and constructive sectors (whose economies appear to be notably decreasing in recent times), respecting at the same time the EU 2020 package, is that of applying a number of policies that create synergy between the various fields of production from agriculture, to commerce, the productive industry, and the building sector, if specific *protocols* were applied in harmony with *bioregionalism* strategy.

The latest term was conceived in the geographical realm to differentiate and classify earth land according to the *bioregions*; that is, the zones in which affinities such as territory, climate, geological stratifications, vegetation presence, and hydrological consistence are stated. Bioregionalism is referred to as

> both a geographical ground and a ground of consciousness to a place and the ideas that have developed about how to live in that place. Within a bio-region the conditions that influence life are similar and these in turn have influenced the human occupancy.
>
> (Berg and Dasmann, 1977)

The bioregion is a geographical zone that presents not only a continuity in terms of hydrology, land-shape, flora, fauna, and other natural characteristics, but also some common aspects regarding culture, constructive traditions, human, and economic resources. A bioregion should be respected and be kept in balance:

> A balance with its region of support through links between human lives, other living things, and the processes of the planet – seasons, weather, water cycles – as revealed by the place itself. It is opposite to a society which *makes a living* through short-term destructive exploitation of land and life.
>
> (Berg and Dasmann, 1977)

The bioregion can also be defined as the better natural organisation of the relationship between the various individuals that cooperate with each other and the intimately integrated and united space which surrounds them (see Devall and Sessions, 1985). The concept of bioregion can become "a holistic way to think about designing a place" (Stevenson, 2013).

The bioregionalist approach for studying landscape or territory requires a number of conceptual initial remarks that will clearly identify its boundaries as well as the other characters, and which are usually employed for differentiating or linking the various territorial areas and for identifying their potential and hazards.

The first concept to be taken into account is that of the *boundary*, no longer identified as a real strict barrier between one region and another, but as a series of alternate filters, buffers, and textures, which are in some way casuals and statistically predictable, but at the same time full of remarkable physical, naturalistic, cultural, and creative opportunities.

As "the final boundaries of a bioregion are best identified by the people who have long lived in it, through human recognition of the realities" (Berg and Dasmann, 1977) then the inhabitants can be considered as an active part of the bioregion itself and should be responsible for the negative as well as positive effects produced by them on it. Respecting a bioregion means to behave in such a way as to operate the choices variably and in accordance with the land and its identity. In fact, provided that "all life on the planet is interconnected, [nonetheless] there is a distinct resonance among living things and factors which influence them, that occurs specifically within each separate place in the planet. Discovering and defining that resonance is a way to describe a bioregion" (Berg and Dasmann, 1977).

Within these boundaries, various aspects are recognizable such as the water basin areas, the flora, the fauna, the cultural, geological and economic issues, and all the reciprocal and dynamical interactions of these categories which act in synergy.

The overlapping of various fields of interparticipation to the local dynamics of the boundaries can be identified by means of a series of different action layers, which are usually linked to the river basin areas, defined in an artificial way, or naturally produced, which are classified according to the resources' influence ray (30 to 50 miles) (see Stevenson, 2013).

At the same time, an approach in harmony with the bioregion's character is beneficial during the moment of redrawing the bioregion's asset, as it allows the guarantee of social and cultural chance for planning in a compromised site (Figure 2.1).

According to the bioregionalist filter, any material in some way incorporates a series of social, cultural, and identity significances of its place, which the bioregionalist decisional process should keep alive by means of adopting specifically local resources.

Innovative approaches to urban regeneration often present similar principles and procedures. For example, according to what Ravetz suggests in his latest book:

> [the] next-generation City [includes] instead of *winner-takes-all* economics and *silo-thinking government* a new model – Urban 3.0 – which is based on creative synergy, networked co-production, and shared intelligence. This kind of synergy can then respond to the complex inter-connecting problems … – climate

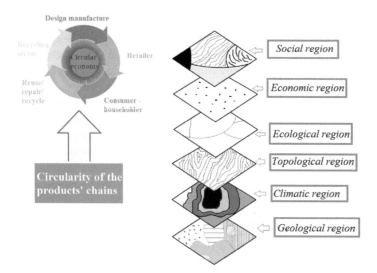

Figure 2.1 Levels of bioregionalism and circular economy

change, social exclusion, economic vulnerability – and then turning problems into opportunities. [The idea comes from the] emerging practice and theory for working with complex problems, and designing intelligent responses. Urban 3.0 uses the method of *relational-emergent mapping*, to explore the opportunities for synergistic collaboration and added value – economic, social, cultural, ecological or others.

(Ravetz, 2014)

The bioregional approach is useful, as long as it can promote the following actions: using materials that have *meaning* in a sustainable manner; keeping alive the subjective and local cultural meanings of materials; respecting the suitability of *green materials* to a particular place; preserving local evolutionary and generative culture of building; and to enhance the *history* of the meaningful local materials (see Stevenson, 2013). "In conclusion a bioregional approach ensures integration of social, economic and environmental factors" (Stevenson, 2013).

b History and definition of bioregionalism

The introduction of the bioregionalism movement in Europe and the comprehension of its cultural evolution could work as an aid to sustainable decisions for city regeneration. The term bioregion comes from the combination of the Greek word *bios*, which means life, and the Latin word *regire*, which means to rule or govern. It is then clear that a region is a very homogeneous geographic territory, in which the rules dictated by nature are more imperative than the laws that people have artificially defined for their proper use and consumption.

"The concept of bioregion has been introduced to explore the potential for developing a relatively non-random method of planning for the wild biological realities of landscape." Bioregions are tentatively defined as biologically significant areas of the earth's surface that can be mapped and discussed as distinct existing patterns of plant, animal, and habitat distributions as related to range patterns and complex cultural niche-habits "attributed to one or more successive occupying populations of the culture-bearing mammal." In particular, many researchers had tried "to discover regional models for new and relatively non arbitrary scales of human activity in relation to that bioregional realities and natural landscape. Equally it is hoped that regional models for an ethic of biotic diversity may emerge" (Van Newkirk, 2009). Sale has proposed, as a deep meaning for the bioregionalism, the definition of nature's government (Sale, 2000). Given the presence of a bioregion, the term bioregionalism has been coined, coming from the geographical studies, for describing the attentive and careful modalities during any action of transformation and/or simplification of the region itself. Bioregionalism is defined as follows:

A decentralized form of human organization, which, by aiming at maintaining the integrity of the biological process, of the life patterns and of specific geographical configurations of the bioregion, can help the material and spiritual development of the human communities.

Rebb (n.d., translated by author)

The birth of bioregionalism, beginning as a cultural movement and then developing as an organized sociopolitical movement, dates from around 1971. During that period, the exact definition of bioregion, as territory that inherits characteristics of cultural and biophysical homogeneity, arose from the collaboration between Van Newkirk and the environmental militant, Peter Berg, one of the fathers of the movement. Van Newkirk had coined his own definition of bioregionalism, which included both the geographic and the consciousness grounds. The idea mainly comes from the need for resuming one's own role within the larger community of the living beings. In addition, by also acting as part of it rather than apart from it, by correcting the behaviors produced by the affirmation of an economic and political global system, which has placed itself out of nature's laws and is devastating simultaneously for nature and human beings. Bioregionalism refers to the ecocentric principles, according to which the ecological balance demands a deep transformation of human behavior. The latest change should take a sensitive role in the planetary ecosystem.

In conclusion, to define bioregionalism, a number of elements and their interaction should be considered, such as "folk, work, place" (Geddes, 2012) and "ecosystems, closed loops, industrial ecology, re-use, re-cycle; local materials and manufacture; a real sense of place" (Stevenson, 2013).

c Bioregionalism in Italy

In Italy, bioregionalism has yet to be widely disseminated and it has only been recognized by a few organizations, including environmental associations, green parties, and the generally left-wing political parties. A little interest has been shown in the anarchic sector and within the ideas of free thinkers such as Massimo Fini, Eduardo Zarelli, Alain de Benoist, and Giacinto Auriti. In particular, bioregionalism has been taken as a principle, which is the idea of employing local materials and the use of building products that have been extracted and manufactured locally.

The benefits derive from the fact that utilizing locally produced and/or manufactured materials (at a county or regional scale), in addition to promoting the development of local enterprises and arts-and-crafts (thus valorizing the character and the peculiarity of the territory), can drastically reduce the pollution levels created by transportation. This is in contrast to consuming a greater amount of fossil fuels—and increasing dependence on oil—generating an exacerbation of costs and emissions of carbon, which are recognized as damaging for the entire planet.

The use of bioregionalism, in such an ancient and historic country as Italy, leads to a number of implications, such as renewed interest in research and knowledge of local culture and traditions, or in the study of construction materials and systems, which are distinctive to the site, even if dismissed and often considered obsolete.

Alternatively, many restrictions for the applicability of the bioregionalist approach to the Italian urban areas can be identified such as the need to verify the local availability of materials and products and/or their chemical and physical compatibility with the construction techniques existing on site.

A number of reviewers and architectural historians, as well as designers, have outlined the importance of recuperating and valorizing site identity within the urban project. Thus, this underlines the opportunity to join such approaches as that of the bioregionalism which cares and takes into account the identity of a place, of a region, and of urban character. For example, Paolo Portoghesi has provided a foundation to his design research based on the notion of place, on the idea that:

> Architecture could not be indifferent to the site in which it arises: from here a theory was born about a space as a system of places...a research that sinks its roots on community values, if not on some aspects of the memory.
>
> (Portoghesi, in Seminario, 2009)

According to the fact that urban areas with high levels of significance can be considered as sites deserving the definition of *place*, which carry with it values, history, human heritage, and richness of landscape, in these areas it is worthwhile approaching any transformation with the bioregionalist methodology.

A place can usually be so-called when and if correlation between the physical territorial characteristics and cultural heritage—that is the people and its traditions—succeed to create a strong identity to the area, so providing the inhabitants and the visitors with a very high level of hosting sensations and of participation.

If, instead, the area is neglected by humans, politicians, and tourists, this means that the goal of creating, by enterprise, a welcoming location has been missed and the identity is lost: this *nonplace* is "a space with no identity, neither producing social relations nor possessing an historical density" (Ilardi, 2007).

Bioregionalism in Italy is strongly related to the cultural identity of a place and it has been in existence for years, among a great number of networks, such as the Italian Bioregional Network.[1] Its main goals are those of a common ground for groups or single people aimed at

> sharing ideas, information, emotions, experiences and projects, so as to develop appropriate life profiles and practices – cultural, social, spiritual, politic and economic – which will be in harmony with his own place, his own bioregion, the other bioregions and the planet.
>
> (Moretti, 2005)

According to this network's ideas, the bioregion, besides being a geographical area in which soil, vegetable, animal species, and climate are recognizable, is also an aggregate of relationships in which human beings are called to live and act as part of the natural communities that define life in the bioregion itself. These homogeneous areas are very often evolved in harmony with all bioregional components.

The network compares the bioregional meaning with the will of renewed patronage, its own bioregion, which should no longer be considered as an entity to exploit, but rather as a whole of beings and living elements, such as plants, animals, mountains, earth, and water, of which humans are only one part. "The social and technological evolution [within this spirit] is ecologically compatible only on a small scale, locally, and if it is anchored to an holistic vision of knowledge" (Moretti, 2005). They affirm that "this consciousness is not something completely new, but roots its origins in the ancient popular knowledge…and in the great western and eastern spiritual traditions" (Moretti, 2005).

The important role played by place is often underlined as essential in the bioregionalist spirit, even though nowadays

> the daily frantic concern and the aseptic and materialistic lifestyle lead more and more to ignore the meaning of the place in which we live, by alienating from the mind any natural manifestation and always longing for a status in continual changing, but never dynamic and creative. [It is not necessary to come back] to the caves, but simply at a lifestyle conscious of the spirit of the place. [The actual role thus becomes that of] living in harmony with the place by means of a balanced, harmonic and deep existence.
>
> (Devall and Sessions, 1985)

It is just this sense of depth that characterizes the Italian movement for the bioregionalism approach, which is strictly connected with the cultural trend of deep ecology. The eight well-known deep ecology points by Devall and Sessions, in 1989, indicate a different way of living in this era. While governments and economists debate around military use of cosmic space, the choice of taking a turn to our bioregions can be considered as an act deeply linked to tradition. "In order to ensure the quality of water, air and food which allow us to survive, we should become our place's guardians" (Berg, 1984).

> The bioregionalism recognizes, feeds, sustains and celebrates the local links: earth, plants, animals, fountains, rivers, lakes, underground waters, oceans, air, family, friends, neighbours, communities, native traditions, indigenous systems of production and commerce. [According to the bioregionalism,] the time has come to know the place's potentials, nature and history, so as to share the aspirations for ensuring a sustainable future.
>
> (Berg, 1984)

Furthermore, the eight points taken to be the essential principles of "deep ecology," are the following (Cobb, 2001: 70):

1. The wellbeing and flourishing of human and nonhuman life on earth have intrinsic value. These values are independent from the usefulness of the nonhuman world for human purposes.
2. Richness and diversity of life forms contribute to the realization of these values and are values in themselves.
3. Humans have no right to reduce this richness and diversity except to satisfy vital needs.
4. The flourishing of human life and cultures is compatible with a substantial decrease of the human population. The flourishing of nonhuman life requires such a decrease.

5. Present human interference with the nonhuman world is excessive and the situation is rapidly worsening.
6. Policies must therefore be changed. These policies affect basic economic, technological, and ideological structures. The resulting state of affairs will be deeply different from the present.
7. The ideological change is mainly that of appreciating life quality rather than adhering to an increasingly high standard of living. There will be a profound awareness of the difference between big and great.
8. Those who subscribe to the foregoing points have an obligation directly or indirectly to try to implement the necessary changes.

Deep ecology sustains the centrality of nature, of Mother Earth and no longer that of humans: mountains, rivers, seas, plants, animals, and human beings have the same rights and the same pride. This leads to the consciousness of being only a part of the complex natural world and to the need for discovering the wild side of our mind.

The critical Italian situation shows the need for diffusing and mainly applying the bioregionalist approach to design procedures, to planning and policies for the city, and to technological and material decisions.

d Bioregionalism and environmental impact

If, according to Heidegger's (2001) words, the *auctor* is who provides an *augere*, that is, an increase, it is nonetheless included in the concept of *augere*, the need for increasing the existing world with a new *thing*, or, in our case, with a new built area. Stating that any action creates an increase, a responsibility is given to the final effects that the previously mentioned action provides, mainly the negative ones, the environmental impacts, the high ecological footprint,[2] and the wide pollution.

It is then desirable to apply an *actio* at reduced dimensions and extent, which can include a minor growth, as the *actio* itself can be identified with its number of impacts; and if, at the same time, the author represents the one who acts responsibly, thus the *de-growth* is without doubt a responsible act of reduction of impacts.

By establishing a relationship between the theory and the practice within the study of pollution questions, defined by aggressive anthropic actions, often the ideas and the theoretical potentials of technical solutions conflict with the physical, the social, and the environmental reality. Therefore, in the practical field, many solutions appear less effective than expected during the planning and design stages. Nonetheless, the application can be generated only by matching the analytic thought with the synthetic one. In fact, the etymology of the word *theory*, derives from the union of the two terms *theo* and *ria*; the first meaning God and the second running. Consequently, the complex and philosophical definition of the term "theory can [be identified with] participation to the Godden event, thus not distance from the entity, but proximity and belonging, to be near" (Aristotle).

If we carry this concept on in the case of city and cultural landscape regeneration, the theory gathers the whole of actions of studying (principles, methodologies, strategies, and solutions): "it represents the moment closer to the question to which we can get" (Aristotle). If the known antithesis between theory and action is seen according to the aforesaid viewpoint, not only can the two terms be configured as two faces of the same entity, but they could also be reconducted to the binomial entity known as project-work. This is done by identifying with, first, the complex of ideas and solutions to a landscape transformation problem, and second, the construction and professional practice.

With Aristotle, we could then "call active at the highest grade, those who play a thought activity" (Aristotle) by identifying as a sole complete action that of *theory/praxis. Praxis* is the activity of "placing the human being between the operating activity and the sensation of being located" and it is finally in this identity that we can find the actual role of the *poiesis*, produced by a poet, by an architect, or by a designer. It is that of being at the same time the one who plays the action and the one who is located in the place, thus knowing the site.

In the human and creative actions a *proaieresis*—a purpose, a choice—can be recognized. Therefore, if the *action* is an increase of objects and things in the world by designers, then the *author*, making their choices in terms of actions, has the duty of applying these latest in a conscious way by either increasing or reducting impacts.

e Bioregionalism and the urban cultural landscape

The sustainable design for urban cultural landscape, or as recently defined, the cityscape, inevitably becomes dependent on the systemic-knowledge approach, which includes both the terms of the technological culture, that is, the users' and the site's requirements and demands. The sustainable-architecture designer can be considered as an interpreter of an existing-urban-area identity, of its natural elements and the local social needs, as a negotiator, and as a deep connoisseur and assessor of the two realms.

In order to accomplish the new alliance with nature, the linear productive-economic system should be eliminated. In fact, according to the latest findings, the whole building process acquires pure resources and elements from the biotic and abiotic realms and transfers their wasted emissions again directly into the realm itself. Instead, assuming a different productive system, which will be integrated into the circular mechanisms of the earthen ecosystems, it is finally possible to reintegrate the anthropic processes within the natural cycles.

The question of bioregionalism and the cultural heritage has recently been merged with the tourism potential of an urban area; regenerating means valorizing, and what more value could be given to a place than that provided by a million people coming from all over the world every year? For example in Italy, where a great amount of the constructed cultural heritage of the world is located, something should be done to efficiently save and maintain the heritage itself. The Italian cities, which need to be regenerated, also require a strong attention to the image that they will show to tourists. Both the bioregionalism and the preservation of the ancient values are to be taken into account and carefully pursued.

The sustainable tourism principles[3] are, in fact, in perfect harmony with both bioregionalism and restoration items. If an area is protected by law and by Heritage

Protection Boards, this is mandatory and any regeneration process, established and switched on by a design solution and procedure, should include the restrictions in the design procedure. As far as all the other parts of the city are concerned, mainly the small areas and the open spaces that are often neglected, the hope of being actually valorized for a tourist destination, could be achieved mainly by employing the right methodology for environmental protection. One of the principal aims is recognized in the need for encouraging the activities of environmental education for schools, citizens, and designers, sustained and supported by institutions, boards, and professional actors.

The regeneration processes for urban areas can assume a strategic role in the safeguarding and valorization of the protected, as well as unprotected, urban zones, thanks to the fact that it can represent a place in which various different aspects of the sustainable management of the city converge. The open spaces in a city are a knotty point in a system of places and paths through which humans can enter into symbiosis with both the natural and the cultural environment with touristic, leisure, or working goals. Giving back to these areas their original significance and value could facilitate, on one hand, the regeneration process and the exploitability by the users. On the other hand, it could create a chance for cultural enrichment and knowledge, with the potential of land control and its heritage presidium to preserve it from uncontrolled and/or speculative interventions.

During the regeneration process of urban areas with interesting cultural heritage, the chance of experimenting with sustainable technologies should be previewed, mainly where the protection law imposes a safeguard and a consideration for existing sites, as sustainable technologies are usually less aggressive, more respectful, and easily reintegrated into the natural cycles.

In European cities, an assorted scenario occurs, but from this differentiation and from the number of difficulties some potentials could be found, as for example, the reuse of

> slums, self-built places, outskirts [as] incubators of social alchemies, [as well as other difficult areas such as the] empty spaces [which can be interpreted as] fields of virtual forces in constant tension between themselves, places of infinite transit where everything passes *hic et nunc*. [The boundary areas of the city can be considered either as a] plan of suspension of decision, [and thus with a negative meaning, or as an element of uncertainty which can lead to an] infinity of possible actions, [and to eventual] production forces of a new spatiality.
>
> (Ilardi, 2007: 43)

At the same time, even the international agencies, such as the Council for European Urbanism (CEU), promote actions aimed at recuperating

> the distinctive characters of European cities, towns, villages and countryside [such as those of] consolidation, renewal and growth by keeping with regional identity and the aspiration of citizens.
>
> (CEU, 2003)

f Bioregionalism in the Mediterranean region

An example of a bioregionalist approach can be found in the ancient architecture and urban development of the Mediterranean basin. Here, the presence of a peculiar Mediterranean architecture has often been demonstrated. In particular, in the literature, the strongest character, common to these areas, has been identified with the bond to the site. In such towns "where all life seemed to live in harmony with its surroundings" (Carson, 2000: 23), the colors, the shapes, and the geometry of the built fabrics show their identity and their desire to be adaptable to the location by employing local materials, climatic connection, and other elements, already present and rooted in the Mediterranean traditional culture. Examples are the bioclimatic response of buildings, the local resource employment, and the social and cultural factors involved in the human activities.

As far as contemporary architecture is concerned, and the new application of city and land configuration, the teaching lectures learned from the previously mentioned principles, included in the Mediterranean tradition, can provide a large and deep aid to the actions and design items aimed at reducing the ecological footprint and respecting the existing landscape, in line with the bioregionalist approach. The word Mediterranean comes from the Latin language and means "in the middle of the land," thereby accentuating the regional and morphologic characters of the area itself, and creating a strong link between those earths that share the same sea and very often the same history. In particular, what is important here to underline is the common architecture language and the regional response to climatic conditions.

"... I see the sea behind the pier and the twinkle of a thousand of sea-flashes. Under my porch goes the wind perfumed of oranges and jasmine" (Annie; a Neapolitan poet). This image reminds us of the construction tradition typical of the Mediterranean regions, which is well known for its choices of climatic answers and adapted to every place with this climate, which is tempered by the sea, mild and warm in winter, sunny and windy in summer. The white walls in the streets are to be defended against the strong sun and to reflect its light. The courtyards act to protect from the wind in winter and employ it in summer, as well as external staircases and arcades; other peculiar devises are the wind towers, the roofs, the ventilation chimneys, and so on. The physiognomy of the settlements is particularly identified, dictated by the functional and environmental reasons in which uniformity and rationality is absent. In which the external envelopes of the fabric of buildings are often variegate and present shapes that suit the land configuration, following peculiar and differentiated choices, which happen to be never standardized or repetitive. These choices were certainly due to arts-and-craft and manual production's needs, but also to the desire of personalizing the homes and interfacing them with the environment.

"The control of nature is a phrase conceived in arrogance, born of the Neanderthal age of biology and philosophy, when it was supposed that nature exists for the convenience of man" (Carson, 2000: 257). This sentence could be taken as a model for the new way of constructing, requalifying, and regenerating every place in the world but in particular in these small regional marine cities.

Some strategies can be defined, which will contain the concept of "plus eco and less ego," by limiting the designer's personal influence, and thus, the footprint. In addition, by recovering the principles of the Western societies cradle, that is, the "mare nostrum" (our sea), by promoting use of dialogue, sharing habits, participation, and reciprocal respect by creating a fan of solutions, not only as far as engineering systems are concerned, but also and specifically of architectural character and of a spatial-configurative matrix.

Examples of these similarities have been discovered in the bioclimatic response of buildings to the weather and environmental conditions, in the modalities of materials employed, and in a great number of social, technical, and cultural human activities (see De Joanna *et al.*, 2012).

Traditional pre-modern spaces develop by searching sites which mainly offer conditions of climatic comfort, thanks to the morphological configuration, to the wise use of materials, vegetation and water, to the exposition to sun or to the protection from cold winds.

(Carbone, 2009: 72)

g Bioregionalism and clean energy issues

The energy question, since its appearance in the early 1960s in the USA, has always been considered as a factor affecting every field of human activity: the building sector, urban development, and the sustainable approach cannot avoid dwelling on this great topic.

The connection between the innovative approach of bioregionalism and the need for saving the consumption of fossil fuels within the entire building process can be discussed starting with legislation. In Europe, as for other law questions, the standards that referred to energy did not always allow a simple application of the fundamental principle. This had pushed towards a new way of managing the relationship between the architecture and energy, that is, the essential requirement of reducing the fossil-fuel consumption in buildings that feed the heating, cooling, and other artificial installations.

The history of construction moved from a model of high urban sustainability—in which shape, technology, poetry, architecture, and performance were joined in order to sustain a structural level of comfort—to a model at zero sustainability. Thus, it is no longer possible to provide security and safety, without the expense of a high rate of energy consumption and environmental pollution. One of the routes can be outlined by finding the fecund, not yet lost, traces of the cultural collective memory, enriched by the fertilization of transcultural processes, that is, the memory of material culture. Humans have inflicted a great number of mistakes and disasters to the territory and landscape during various civilizations. This was not due to the will of damaging and surmounting the ecosystems or to a desire to build an unhealthy and hazardous habitat for users. Nevertheless, in this absentminded and unsympathetic way, any human means of land transformation provided both to the environment and to living beings can no longer be endured, as the decay now over the resilience threshold created by a lack of attention and care, is evident.

> The human settlement, with its physical and socio-economical arrangement, is today a terrestrial family which aims, and in part has already succeeded, to develop models for life, consumption and use of resources, as if the territorial restrictions, with which he is obliged to interact in the long time, did not exist. Historically the restrictions which linked the settlement to its environmental surrounding led the community to project their actions within themselves and its milieu of reference. The introduction, or better, the rediscovery of the renewable energy sources is drawn as a chance for developing interaction shapes which will be *sin-energetic* with the land and projected again to its reference environment in a (self) sustained way.
>
> (Borghini and Tatavitto, 2012: 339)

This means that according to a bioregional approach, clean energy can be produced only if the systems, the plants, and the prime-energy-sources are found locally.

"Using new technologies for the renewables, without those deep relationships with local region, which are able to transform the nature's elements into resource, could lead to look for energy inputs out of the territorial context of reference" (Borghini and Tatavitto, 2012: 340). Thus, out of the bioregion, excluding all the advantages due to the local resources' exploitation. "From the energy viewpoint, the increasing dependence of a settlement on its local environment involves the use of external resources…which look towards a growing disconnection between the settled society and the territory" (Borghini and Tatavitto, 2012: 340).

Starting from the hypothesis that any construction and urban open space dedicated to human activity requires—during its whole lifecycle, from the programming and application of the design stage until its discharge, through the stages of siting, use, maintenance, and management—the employment of a certain amount of energy under various shapes. Given the fact that this energy could be produced by a number of various sources, the attention could be focused on a primary aspect. The amount and frequency of this energy-use depends on serial factors, numberless and interrelated among each other, such as the architectural, spatial, constructive, technical, planting, material, and management elements that are present in the urban area itself.

Therefore, by isolating the energy-saving question only to the completion stage on site of the already built goods, or furthermore taking into account—as a unique factor of fuel waste—the selection of artificial plants and systems does not appear only as limitative, but disadvantageous in terms of the achievement of the goal that the law and all other policies for energy saving establish. This goal can be identified with the need for creating a positive balance between the collected resources, the employed resources and the wasted ones, which will allow an almost unlimited length of raw materials without making inroads into welfare levels. It is very important to "merge the energy resources potentially available in the territory with an appropriate urban design" (Vanoli, 2012: 321).

In the past, the situation was different. Since the first human settlements, energy, as well as human habitat questions, have been braided into each other and they had led to the invention and construction of peculiar devices such as arcades, covering streets, and green squares with trees. These were aimed, on the one hand, at improving the viability conditions for the inhabitants and, on the other hand, at reducing management expenses of life itself. But, generally speaking, the whole architecture of the settlement, either by putting the buildings very near each other, or by leaving some space between them (according to the environmental existing factors) generated an organism able to match the environmental and climatic inconveniences and to favor better conditions for human comfort.

> Often the urban traces appeared as a prolongation of the natural order, in a logic of complex harmony whose results can still be admired now. [It is fundamental today to affirm again the centrality of the climate] in order to re-establish that direct relationship with the natural environment which now seems lost, overwhelmed by modern over-structures ... which take away the personality of a place and cut off more and more the product of architecture of man and its living.
> (Carbone, 2009: 72)

An energy-saving approach to urban regeneration should consider and analyze the peculiar phenomena that occur within a city, which can help to create comfort conditions, or on the contrary, to provide disadvantages and discomfort to users. The knowledge acquired by means of deep and scientific investigation of the heat paths in the city is the first tool for applying an energy-conscious design that will save fossil fuels by including technological solutions that will employ natural resources as means of providing clean energy.

One of the important aspects, discovered during studies on urban open spaces, is defined as:

> [The] so-called Urban Island Effect, which depends on how the city is built, on the employed materials, on the drawings of streets and buildings, besides on the artificial heat generated by the hydrocarbons' combustion for transportation and domestic use. The streets' and buildings' geometry itself, due to their shape and height [which create the canyon effect], as a mirrors' game, traps the heat in between them, before releasing it in the atmosphere.
> (Selicato and Cardinale, 2012: 323)

The temperatures within the city are thus subjected to variation due to the concentration and quality of built areas. "The urban morphology, i.e. the tri-dimensional asset of buildings and urban open spaces, plays a fundamental role as far as external microclimate is concerned" (Selicato and Cardinale, 2012: 325) and leads to various microclimatic conditions. The following is even possible:

> To address the choices at urban scale towards a sustainable transformation of free areas into zones for clean energy production. In some districts there

are examples of an important effort for preparing the city for the growth, for fighting and adapting to the climatic changes, for improving the life quality for citizens...such as the Chicago Climate Action Plan or the Greener Greater New York, defining a development strategy which includes the energy sustainability within the more global environmental sustainability. In fact the development lines for the contemporary city cannot avoid being compared with the questions due to the climatic change.

(Marotta and Schilleci, 2012: 548)

Under the light of the environmental question, the planetary pollution of water, air and soil, climate change due to the greenhouse effect of overheating the earth, and the feared end of oil sources that feed 85 percent of our consumption, the relationship between energy and architecture is finally confronted in a more holistic way. New conceptions emerge aimed at creating a balance between the built and natural environment by meeting the new as well as the common users' needs. In addition, innovative ideas are aimed at integrating the construction portion of the urban areas with the landscape, climate, and preexisting cultural heritage, which could favor both high levels of comfort and a great saving of energy sources.

The application of more natural processes for conditioning, which even in the open spaces of a city could replace traditional and polluting ones (the gas or gasoline stoves, for example, used for the open restaurants), has already been under the attention of scientific researchers at MIT since the 1960s. These systems seem to propose a partial solution to fossil fuel consumption and to consequential emission of contaminating, carcinogenic, and atmospherically modifying substances. The question has to be approached from fundamental principles, that is, from the conception itself of architecture: space selection, plan disposition and height towards the sun, wind, and land, technologies to be employed, construction materials, and management systems; only in this way can the question of resource savings on the planet and their durability be faced and responded to in a correct and complete modality in order to contribute to sustainable development. Some superficial habits should be avoided, such as that often operating in Italy, where "the standards on energy saving are applied only by increasing the thickness of fossil-made insulation systems on walls" (Allen, 2008: 2). Furthermore, the materials employed for increasing this thickness are frequently neither sustainable nor healthy, such as polyurethane or polystyrene.

In conclusion, it can be said that the connection between bioregionalism and energy issues can be identified with the fact that the bioregionalist approach promotes the use of local and natural resources for satisfying primary needs, which also requires the use of renewable sources of energy (see Berg, 1984).

h Bioregionalism and low-tech solutions

For a long time, research has been focused on the creation and design of hi-tech components for buildings with a high performance level in any kind of situation to promote a global scale dissemination and ensure large market segments and diverse economic income. This logic, based on the increase of goods' production rather than on saving resources, has provided, through time, a number of negative effects. For example, the exploitation (if not a complete depletion) of raw materials and pure energy sources and a standardization of architectural language and a cultural flattening by proposing stereotypes and formal layouts that were insensitive to site diversity and peculiarity.

Within recent decades, a tendency has been observed towards importing to other regions (mainly in the Mediterranean or in the southern world countries) tastes and fashions from northern and western regions which, confident of their dominant position in the global market, propose and impose in any field a number of models that affect desires, aspirations, and lifestyles in the remaining part of the planet. This globalization phenomenon is clearly shown in architecture and additionally leads to a hazardous process of cultural weakening and morphological homologation that tends to transform cities into a common empty standard without any meaning. All this has had a strong negative effect on the natural environment for a number of reasons.[4] The first reason is that northern models require the help of foreign materials and techniques, very often produced and found in areas very distant from the bioregion of the city itself, thus polluting and not belonging to the local ecosystem. The proposed design solutions are not appropriate to the geographical, climatic, and cultural conditions. A different model of designing contemporary cities in any specific region, and in particular within the Mediterranean area, should take into account the important role of materials as a decision item within the architectural process, affecting both the shape and the contents of the final construction.

Studies and reprocessing of traditional techniques allows an efficient exploitation of available resources by strategically achieving sustainability goals, and at the same time, respecting the cultural and technical expertise and wisdom of the local regional conditions. In recent years, interest in vernacular construction has been emerging. The use of traditional constructive systems is increasing, such as the use of stone and adobe masonry, timber construction, and mixed building techniques using a combination of timber with adobe masonry. Even the use of industrialized materials does not imply that architecture becomes insensitive to the references of popular architecture, but on the contrary the contemporary architecture can learn from the past practice.

Today, a number of technical solutions are available in technological expertise and sometimes already in the building market which, learning from the past, could guarantee high levels of performance, even limiting the use of pure energy and raw materials and the production of solid, liquid, and gas wastes. Put simply, low-tech, of low environmental impact and reduced ecological footprint. The design criteria for these strongly humanized and historically great areas should take into account the influence of changed cultural awareness that characterizes the spirit of the past and considers the significance of its attempts to reinterpret the past itself. It is also necessary to review the relationship between architecture and landscape as a response to both environmental conditions (climate, topography, water sources, and preexisting structures) and cultural expectations (the villa in the country, intellectual priorities, and application of technology), as well as to be familiar with the use of water, circulation, architectural disposition, and organisational patterns. Some examples of low-tech solutions will be described in parts II and III.

The concept of bioregionalism is based on the selection of materials and it should not be antagonistic to other environmental agents involved in the design process. On the contrary, according to this approach, indigenous materials and products should be encouraged to reduce energy costs as well as transportation pollution.

i Participating design (bioregionalism of people)

The participation topic within the decision and programming stages of urban regeneration, which is recognized worldwide as a current question, is strongly debated within the new social and territorial policies and it should obviously be included in studies about sustainable architecture and technologies. The UNESCO definition for the participation programs declares that "the participation is a process of re-belonging of power, the development of people's individual and collective capacities of improving their own existence as well as of conquering an increasing control over their own destiny" (AAVV, 1999: 21).

As far as architecture is concerned, the participation process occurs when there is a presence in the design stage, completion stage, and in the management stage of the customers. It is also fundamental that the building's user understands the best modality of usage (see De Lucchi and Pastenga, 2008: 9). A participated project should activate sharing strategies by involving the citizens to make them work upon a common idea aimed at a balanced and sustainable urban development. A participation design is aimed at social groups' advantages, starting from their actual needs, and involving the local inhabitants in an active way. "A project, which makes the citizens able to actively participate, is mainly oriented towards the process rather than to the product" (AAVV, 1999: 20). The process is what makes the difference between a regeneration design procedure and the conventional requalification and/or reconstruction path.

> The participation process, less linear, more complex and articulated as far as the social and psychological items are concerned, requires continuous re-arrangements of the definition of the problem based on the experience and on the knowledge achieved by any process step, by the citizens' active collaboration.
>
> (AAVV, 1999: 20)

A project governed by the participation procedure applies various approaches and strategies, as far as both technical and procedural items are concerned. "Such a project becomes enriched by the concrete functional response adopted for satisfying the needs, both as single and as community, emerged during the procedure between the involved citizens, the technicians and the experts" (AAVV, 1999: 21).

Participation can become a part of the city's regeneration procedure. Apart from the necessary analogy between social sustainability and environmental sustainability, it is important to recognize how the participatory approach to design requires knowledge of local processes. The architectural design is based on sharing of those choices, in terms of space, shape, construction, and dimension, thus resulting in being agreeable, comfortable and functional for the future users, who should participate within the selection of the elements of the built settlement. It is mainly in the relationship with the place, as an anthropic process, that the role of participation design is played, as the regeneration processes are not exonerated by this need. Conversely, they can take great benefit and derive strong positive contributions from the relationship itself.

Many researchers already agree on the fact that participation should be included in the bioregionalist approach to urban regeneration. The well-known architect Richard Rogers stated: "design…is key to (the) achievement of *quality*" and the design itself should provide "special attention to the needs, the aspirations and the experiences of urban inhabitants instead of the kind of aims for superficial aesthetic goals" (Thwaites *et al.*, 2007: 4). If participation during the construction process was finally considered as a social and simultaneously cultural and fundamental attribute of a project, mainly for the regeneration of cities, it is also very clear how bioregionalism can actually become the perfect approach to such a policy.

> The bioregional perspective creates again a feeling of participation with the local identity, grounded on a renewed critical consciousness and on the respect for the integrity of our ecological communities…The life security starts [in fact] with the assumption of responsibilities at local level.
>
> (Berg, 1984)

How participation can affect the social improvement of a city can be synthesized in the following poetic document:

> I had a dream. A dream within which the City was a livable place and whose maintenance and care became chances for participation, for citizenship exercise, for occupancy of plural junior employees, and where the work was a tool for social inclusion. The citizen was giving his own contribution of ideas and processing in the identification of needs and in the research of solutions; life quality and health in the places were the bases for the human actions; the technicians together with the citizens were designing a livable city and were programming use and maintenance; care and maintenance needs became a basin for positioning and chance for exercising the participation; the young in strict relationship with the citizen, the technician and the enterprises learned an art and found a position; the work became a tool for release and socialization for less advantaged persons; children lived in a more healthy context.
>
> (Tanese, 1999: 5)

It is clear that the main goals to be achieved by participatory action during the design stage are those of "creating a natural and sustainable environment of optimum life and establishing such a modality for being and inhabiting, able to build and re-build social textures and human relationships which we normally mean generally as *citizenship's rights*" (AAVV, 1999: 11).

In the field of architectural works, the more socially involved architects of South American countries confirm the importance of social goals. Aravena recently declared that "it is not a bit unusual to take architecture as a socio-political subject: not many people see it in that way. Nowadays an architect can choose to stay apart from this viewpoint. We have selected a different road" (Aravena, 2009: 109). It is clear that any "social programming requires more and more often the construction of participated projects" (De Ambrogio, 2009: 1). The history of such participated projects is a tradition from the last century.

> The subject of the participation within the design culture at the various scales is a recurrent question and lately has been assuming an increasing interest. The problem of interfacing the design stage – i.e. the decisions regarding the social-demand issues, at urban level, for services and facilities aimed at satisfying the individual and collective needs – with the users' as well as consumers' expression of desires and inclinations has long been the core of the debate and of the public initiatives for programming and planning the interventions.
>
> (Schiaffonati and Riva, 2014: 101)

For example, the *Contratti di Quartiere* (neighborhood contracts) allowed the citizens' participation in local authority decisions and guaranteed a better tendency for compensating social unbalances.

> In the Italian context the question is set in the 50s with the return of democracy, within the frame of hopes and expectations of the post-war reconstruction. A process of high political participation occurred not only with the urban and fabrics' reconstruction, but also for the re-birth of the civil institution, of the land authorities [and so on].
>
> (Schiaffonati and Riva, 2014: 101–2)

> [In the 1960s,] a number of district committees developed in almost every Italian town, as a spontaneous structure of participation in line with the claims and fights for the *right for the city* and for the application of an administrative decentralization able to approximate the citizens to process and share the municipality's choices.
>
> (Schiaffonati and Riva, 2014: 106)

The district inhabitants' role was that of being invited to "illustrate their world by means of a drawing which describes how their district works according to their life experiences" (AAVV, 1999: 21). We all know that "it is not enough to put the persons around a table to produce magically a good participated design" (Balducci, 1991).

Today, after the end of the fashion phase of the approach (which characterized the end of the 1990s and the first new millennium years) "the participated design has to be considered as ineludibly since it is fundamental for constructing social projects...not for ideological or value reasons, but essentially methodologically and with contents" (De Ambrogio, 2009). Given that participation means collaboration between various parts of the same community at the same time and in the same space, the idea of a relationship between the participating projects and bioregionalism resides in the harmony created between the public arrangement of a settlement and the level of possible utilization of any shape of energy, matter, and object in the space: this harmony leads to the new concept of *common good*. A city is a perfect entity that defines the place for every typology of common goods, from the more material to the less concrete ones such as air, water, and energy. The definition of *common goods*, as processed by the Rodotà Commission, is the following: "goods of collective ownership, out of the market logics and of the profit, aimed at the satisfaction of the fundamental rights."[5]

The famous idiom by Don Milani, *I care*, which has been taken as a symbol for the habit that can provide wishes for participation and as part of the feeling of community, has nowadays been transformed into the expression: *it is worthwhile for me* (see Milani, 1965); for example in southern Italy:

> In a perverse way, the practice of *it is worthwhile for me*...leads to the abuse of the common good, to the disdain and to the deceit towards the others... which mainly hit the weakest individuals, such as the unemployed, the elders, the young, [the foreigners] and all the emarginated persons, to whom a second-hand citizenship is assigned.
>
> (AAVV, 1999: 12)

According to the rights established by the UN, among those aimed at satisfying the primary needs of a population, there is also food, water, and air. Thus, the city should be considered as a store for any citizen that would like to satisfy their needs. Although nowadays food and water are purchased, it can still be considered that the other services, which do not require great amounts of energy and labor could be distributed without any expense. The more important issue is quality of life: this should be guaranteed in any city, especially in their open spaces, which are places without any private property and thus could be defined as common goods. The participation process to any modification of the urban spaces, is thus not only eventual or optional, it is the only possible procedure to be applied to guarantee both human bioregionalism and the ecosystem's bioregionalism. The urban context in which any research is included affects the transformation, and thus, should be taken in great consideration, in particular, as far as the contextual system of people is concerned (Carabelli *et al.*, 2011: 6). An urban sustainable regeneration is then the right procedure "with the perspective of orienting the territorial and building policies towards a wider sharing and participation, for the administrative institutions' growth" (Schiaffonati and Riva, 2014: 101–2).

The design action should go beyond the rational settings, which require a strong control of actors' behavior, and assuming incremental logics, within which the project is established, evaluated and corrected in operation (De Ambrogio, 2009: 1).

> It is essential…to plan together the project and the consent, and to build the sense of belonging within the participants as far as ideas and paths to be applied are concerned, so that the choices could be effective and applied without meeting great resistance and opposition.
>
> (De Ambrogio, 2009: 1)

In order to develop the participated design as a fundamental element of the bioregionalist approach, a number of actions should be applied. These include providing greater accessibility to all kinds of social bands, with particular care to those less able, and identifying green areas integrated with a number of facilities that could favor physical activity and at the same time social integration and conviviality, with a definition of alternative parking areas. In addition, urban furniture should be designed with originality and ecosustainable aims: sitting areas, flower buns, lighting systems, and so on, created with recycled and/or recyclable material.[6] Furthermore, the potential of the site for economic and social requalification should be taken into account through the promotion of economic and social activities, which the regeneration design could activate and favor.

Another important aspect of the participation approach to urban bioregionalist regeneration is that of the preliminary study. In particular, some investigations include analyzing the evolution of the public actors' localization as far as regeneration and sustainable development is concerned. Other investigations include creating an intelligibility of the action system (that is, the relationships that link the actors able to transform the city), allowing comprehension of the functionality of the main aspects of sustainable development and for the valorization of the eventual existing traditional heritage; the two final studies to be carried out for participatory process are previewing a scenario of regulations for the action within the specific frame of the consolidated boroughs, and progressively building a database of best-practice samples (see Carabelli et al., 2011: 6).

One of the recognized methods for applying a participatory procedure during the urban regeneration project is the "ten-step path" (see De Ambrogio, 2009: 2):

1. Definition of the first concept.
2. Actors' identification and involvement, and agreement's establishment and activation.
3. Meta-project definition.
4. First analysis of the context and need.

5. First meeting of the project team with the common question definition.
6. Second meeting with definition of the alternative strategies for processing the question.
7. Third meeting for building agreement according to the proposals.
8. Fourth meeting for designing the projects' completion step.
9. Fifth meeting with establishment of the self-assessing procedure.
10. Sixth and last meeting for building and for analyzing the budget and the whole project's revision.

Thus, participatory action research can be presented as an anthropic methodology and process. It is a goal that any social study should always strive to achieve. Its role is that of promoting social change through organisational learning; participatory actions can often work as a link between cooperation, social action, and knowledge generation (see Whyte, 1991).

Environmental design, born following performance design and the ecological movement, does approve of the bioregionalist approach and methodologies that try to safeguard the cultural as well as the natural heritage. Whenever the need arises to transform an urban settlement, the bioregionalist vision can help to take into account the various levels of the bioregion under analysis. At the same time, it reduces impact, valorizes local richness of landscape, and exploits natural energy as well as other resource use, while adopting sustainable technologies and people involvement.

Notes

1 The Italian Bioregional Network was founded by Giuseppe Moretti in the spring of 1996.
2 This indicator, the ecological footprint (EF), created by Mathis Wackernagel a number of years ago, can measure the level of human's impact on earth. Given, in fact, that the available amount of resources on earth (soil, food, material, energy, and water), that is, the biocapacity (BC), has been calculated as 1.8 hectares per person, the countries that have an EF superior than BC are actually debtors, while the countries that create an EF inferior than the BC are actually creditors. Most of the countries have overcome this limit, are debtors and are employing more than it is possible: from here, there is a need for embracing new roads of development.
3 Sustainable tourism has the fundamental aim of attracting the attention of a curious and careful demand to the material and immaterial culture of places. It favors building regeneration to adapt to new requirements and of valorizing the existing built heritage, both the precious and the disused, mainly those with the spirit of the place, by limiting the construction of new fabrics and any environmental impact. In addition, reducing the use of primary resources and favoring the economic convenience for the long-term. Many organizations have been keen on this question lately. For example, the World Tourist Organization (WTO) established a world code of tourism ethics in which it defines tourism as a factor of sustainable development. It proposes to overcome a sole economic declination, and instead, orienting towards a durability of resources of the hosting territories. It proposes to users, operators, and local societies, of a more careful confrontation upon the subjects of cultural respect, the environment, and the market economy. Other panels, such as the UN meeting in Johannesburg, with the responsible

tourism act in 2002, or the sustainable tourism WTO and UNEP United Nations environmental program states that:

The sustainable tourism is what generates an optimal utilization of the environmental resources, as key elements of the tourism; when the essential ecological processes should be maintained, where the natural resources and biodiversity should be saved; where the historic-cultural identity of the hosting communities should be respected, by saving the typical constructions, the cultural milieu and the traditional values; and finally where the tourism should ensure long-term economic actions, by providing socio-economic benefits, widely distributed, among which: a stable employment, budget opportunities, social facilities for the hosting communities.

4 For the link between globalization and architecture see Chapter 3a.
5 This definition came out from the Rodotà Commission, engaged by the Prodi Government (2007 to 2008) for modifying the civil code in the part referring to the public property (see Lucarelli, 2011).
6 For the issues related to recycling, see Chapter 4.

3 Zero kilometer materials and products suitable for urban regeneration

a Local and global economy: the conception of zero kilometer processes and supply chains

The interaction between globalization and the consumerist society is one of the main issues that the present era faces in order to comprehend the processes of anthropic evolution in the third millennium. Therefore, every possible effort should be made in avoiding negative effects on the weakest groups and exploiting the benefits out of any human activity towards both the human and other inhabitants of the earth as much as possible. In particular, the building process, changed according to the previously mentioned statements, and the city regeneration subject, focused on here, are also affected by these big changes. In fact, among all the *green* movements and approaches to the new economy, the conception of zero kilometer (km) materials has become, by now, one of the main goals to be achieved with any sustainable urban policy.

As far as the architecture of the city is concerned, the effects of globalization are evident, and they can be clarified by a brief description of the phenomenon and its impacts on the building sector. It is known that "in 2008 the world population who resides in the city was more than half of the total population" (Moccia, 2012: 183). However:

> In the territories at strong urbanization the quality of landscape is disconnected: in the new desired and undesired city the new worldly GDP (gross domestic product) will be increasingly produced with various levels of territorial and social inequity…Is it then possible to work on imagining a new transition of these areas towards *the other city* which will help the fragmentary space to be reconnected, to be darned till being recognized as a more sober city, with different values, more careful towards the environment and to the relational values?
>
> (Persico, 2013: 11)

One of the paradigms of this millennium can be encompassed in the Pliny Fisk's sentence: "disassembly and reassembly must be the wave of the future" (*Texas Monthly*, 2008). Possible links between the waste issue on our planet and the role of architecture within the process of reconstruction for a participatory future needs to be conscious that, on one hand, architecture and land transformation should abandon the idea of a consumerist and commercialized society (which had led to the application of bad practices). However, on the other hand, a sustainable management of the artificiality can contribute to reducing the excessive mechanization without renouncing its benefits and any new discovery.

The twentieth century had focused on technological progress and a part (unfortunately too reduced) of humanity had benefited from this progress. Paolo Soleri had identified six principles, corresponding to an equal number of human benefits. In the twenty-first century, it is possible to ascertain the contrast created between Soleri's principles and the present development. Growth and technological progress should have provided welfare for all of humanity, who, instead, suffer for the collateral effects of this progress, which, more recently, has demonstrated disrespect towards the planet. "The various efforts to reshape the vague materialism in a reflected balance, where the production-consumption value set against an elegant and illuminated trinity, which works on the base of knowledge, apprehension and transcendence" (Soleri, 2009) had led to reducing the damaging effects of artificiality by means of employing nonadulterated materials. Alternatively, the solution lies in leaving the things in a state closer to their *natural* condition to minimize the toxicity effects by reducing or completely avoiding the high-tech production methods that consume high amounts of energy and release polluting substances. One of the possible routes to follow is to define, for each stage of the building process, which actions can be carried out by entering nature's cyclical course, for assembling and disassembling the matter in order to make it always pure and usable, and to match the principles of sustainable development.

It is useful to compare the current situation with *globalization*, which is, according to Habermas (2006: 175):

> The cumulative processes of a worldwide expansion of trade and production, of commodity and financial markets, of fashions, of the media and computer programmes, news and communication networks, transportation systems and flows of migration, the risk engendered by large-scale technology, environmental damage and epidemics, as well as organized crime and terrorism.

This new way of organizing human activities presents an immense dominium of diffusion, whose concept can be clarified by the definition: "globalisation is... more complex than the simple expansion of Western capital and concomitant spread of products, culture and style" (Habermas, 2006: 76). In fact, it can be said that it "is a new World-Order. We do not know its outcome or have a full practice of its nature as we are only in its earlier stages" (Habermas, 2006: 47). However, a number of effects, either as benefits or as negative impacts, are to be shown very clearly at this stage of contemporary culture. For example:

> In architecture, the historic development of globalization corresponded very closely to the ascendency of Modernism. Founding Modernism ideals had always been global in ambition...For countries swept up in the tide of global

economy, the association of Modernism with rationality, progress and successful and dominant North-Atlantic economics was irresistible.

(Habermas, 2006: 74–5)

According to the philosopher's words, modern architecture does not go in the
direction of a sustainable development of a city, if globalization means trying to
make the various locations of the earth similar in terms of image, architectural
strategies, technical solutions, and citizens' social behavior. If citizens intend to
achieve the same goals and aspirations all over the world, the risk is very clear:
losing local identities and the peculiar characteristics of any material and cultural
value. This can surely be a negative consequence of globalization, which is creating "in a very short space of time the *homogenization* of global consumerism."
One of the greatest hazards of globalization can thus be recognized in the "*homogenization* of city centres throughout the world" (Habermas, 2006: 75), which runs
in parallel to consumerist strategy.

Consequently, architecture is uniformed by the star-system, within which, if
"the intention is that the building should be an *iconic global product*, then local
distinctiveness is often not a desirable character" (Habermas, 2006: 75). In fact, it
has become very clear by now that "this interest in local contextual identity runs
contrary to the homogeneous international architecture of star architects and their
followers" (Adam, 2008: 77).

Indeed, even among the most famous and clever professional architects and
designers, it is today possible to find two very different approaches towards the
cultural environment transformation: these "*two poles* of modern architecture –
super-modernism and *the particularity of place* – are clear reflections of the two
poles of globalisation – *homogenisation and localisation*" (Adam, 2008: 77). It is
then comprehensible that for the architecture of the star-system:

> The *surroundings* constitute neither legitimization nor inspiration; in fact
> these are derived from what goes on inside the building, from the programme.
> This autonomy is in many cases reinforced by the fact that the building has
> an inscrutable exterior that betrays nothing of what happens inside…in many
> instances these buildings look as if they might house just about anything:
> an office or a bank or a school, or a research centre, a hotel or apartments, a
> shopping mall or an airport terminal.
>
> (Ibelings, 1998: 88)

Moreover, "the reproduction of the spiral or the twisted forms, globular glass,
planar intersection (and so on) is facilitated by the use of the same sophisticated
computer graphics employed by the offices of the star architects to develop and
present their concepts" (Habermas, 2006: 75). Thus, reproducing a never-ending
similarity for any site, any location, any different situation, and for the population's requirements.

What kind of comfort, homeliness, and hosting context can a city create when
there will no longer be identity and differentiation between one and another?

How can the culture, traditions, and the peculiarities of a population inhabiting a city represent a value and strength, if the cities are the same all over the world? Which kind of biodiversity can be found in the planet, when everyone is living in the same kind of building, in the same shape of squares and streets and green parks?

Alternatively, it is known that globalization has a number of qualities that have provided and will continue to provide benefits to both citizens and cities. For example, the possibility of interacting with each other through information systems, the chance of sharing similar adventures and ideas, and the great advantage of talking easily and quickly.

The consequent benefits for the architecture in the city can be found in the reorganization of life and in the recognition of local peculiarities, aimed at sharing the cultural and traditional values with other communities by taking advantage of the boundary effect of intercultural society. Within this vision, urban regeneration could contribute to the efforts for overcoming decay on the outskirts and any social debt questions.

The visibility of the problems and the research for the most suitable solutions to the urban and social contexts are the first issues to be addressed. According to the vision of illustrious scientists such as Paul Connett, the solution to these questions are to be found in the complexity of a "planned *de-growth*, [which will finally lead] towards a sustainable steady-state [situation. At the same time, in order] to meet the needs of a growing world population." He proposes a strategy in which the "industrialized-world's reductions in material consumption, energy use, and environmental degradation [will be] over ninety percent by 2040." As far as the global temperature increase is concerned, he said that to avoid this effect, "the world must reduce carbon emissions by ninety percent by 2050" (Connett, 2013). Thus, "OECD nations should be taking steps to reduce their ecological footprints of a range from fifty percent to eighty percent" (see Rees, 1992; WWF, 2006). We can also say with Gary Gardner's words:

> In an era of rapidly rising demand for materials, nations would do well to institute policies designed to conserve non-renewable resources. One solution to the problem is to reduce radically the waste of non-renewable resources through adoption of policies that promote *circular economies*, in which materials are circulated over and over rather than being sent to landfills.
>
> (Gardner and Prugh, 2008)

By now, even though conscious of the environmental and polluting problems on our earth, "...people say they think we should pollute less, but they also intend to continue to do things which pollute (driving, flying and so on)..." (Finch, 2008: 27) and do not think that it is a personal challenge to contribute in synergy, each one within his small piece of work and life habit, to reduce the environmental hazard. The need arises to look for a stronger political change:

> In a consumeristic culture, desegregated behaviours and desires are held to be paramount unless decision-makers and their supporters can define, establish

and defend robustly a notion of greater good. Consumeristic ideology means worrying that airlines may be overcharging rather than aviation fuel should attract taxes. *I shop therefore I am*...even on the internet.

(Finch, 2008: 27)

A fundamental to our society has been established:

> The richness cannot be originated by the job which requires time. With the speediness of time, the finance had replaced the industry, and the money has replaced the time. With the globalization, money has become the sole value. As a consequence the Latin idiom – *homo homini lupus* – has become a reality. It is bad that it has been said that who has money can eat; but this is the core, the essential requirement of our western culture.

(Galimberti, 2013)

However:

> The satisfaction a person obtains from his income does not depend on its absolute level but on its relation to the others in the same community at the same time...If there is an increase in the level of income with no change in people's relative position, then nobody feels better off.

(Abramovitz, 1979: 8)

As "globalization is a quite recent phenomenon spanning the last twenty years, or less, of economic development" (Santarelli and Figini, 2004: 2), then "the current use of the term globalization soon recalls never-ending horizons and deep phenomena, in any activity's fields. And still the word, since few years ago, did not show any familiarity even with the Anglo-Saxon world" (Mattoscio, 1999: 24). The following has been noted in globalization:

> [It] can be defined as a historic process driven by: technological factors, such as the development of computers and the Internet, which reduce the distance between people in terms of both space and time; political factors, namely the demise of the former communist bloc of countries, which meant the end of one of the two systems of production and allocation of resources historically determined: the centrally planned economy; economic factors, partially as a consequence of technology, which have led the *global world* to adopt free-market oriented economic policies and individual behaviours. As said, globalization is a multifaceted process – characterized by a wave of privatization in public utilities and other previously state-owned industries, reform of both domestic financial markets and taxation systems, and liberalization of labour markets – which has produced unprecedented acceleration in the flows of both international trade and Foreign Direct Investment (FDI).

(Santarelli and Figini, 2004: 7)

"A third factor characterizing globalization is privatization" (Santarelli and Figini, 2004: 15). In fact, "the diffuse opinion is that the world economy not only has overcome the national spaces, but is changing its aims from the international to the globalized ones with a dynamic...which can be defined as inter-globalization" (Curzio, 1999: 202). The government as well as any other transformation operator, including the engineers, the designers, and the planners, should take the role of creators as well as maintainers of the built environment. They have the need to take responsibility, they have to "try to find a common design language, a common analytical method, [and also] should push political decision-makers to take on the big *issues*. A level regulatory playing field [is needed, which will act] for the community's benefits" (Finch, 2008: 27).

The city architecture has its big role and it can help during this process of societal goals. In fact, "architecture brings together...[what]...is important for society at large: shelter, social function, technology, art, economics, politics, science and more" and if the architecture can be a *mirror* of society, also "the view can be reversed: society can be the mirror to architecture" (Habermas, 2006: 75). Thus, creating the need for great exchange between the two entities: the social and the constructed worlds.

"The creation of the *rights* of the individual over their resident community or state had damaged the role of the relationships between the state and the individual" (Habermas, 2006: 76). The important concept of *localization*, which stands as the other face of globalization and which can create an alternative and/or a synergy, "is closely associated with the politics of identity. Identity is community and place-related, and the individuality of community and place are undermined by global homogenization. Migration, instantaneous communication and increased travel all threaten and dislocate community identity" (Habermas, 2006: 76). What is the connection between globalization, localization, and regionalism within the city architecture establishment? In the past, a building could be considered *regional*, when it was vestigial and it had a strong link with the external context. However, at the same time, this character should not be "sentimentally identified with the vernacular" (Frampton, 1987: 20–7).

Standing opposite to this concept of regionalism, recently "the quality of place, a city or region, has replaced access as the pivotal point of competitive advantage; quality-of-place features attractive to talented workers of a region have thus become central to regional strategies for developing high-tech industries" (Habermas, 2006: 76). During regeneration processes in the city, it is necessary to "favour cultural differences, every city's own identity, and...believe that city planning and architecture should emphasize these proper symbols" (Borja and Castells, 1997: 227–30).

In 1993, the Congress for the New Urbanism were looking for "the reconfiguration of sprawling suburbs into communicative areas of real neighborhoods and diverse districts...and for the development of towns and cities [so as to] respect historical patterns, precedents and boundaries" (Congress for the New Urbanism, 1993). However, we had to wait until the third millennium for implementing real applicative reformations that could promote actions aimed at safeguarding "the distinctive characters of European cities, towns, villages and countryside…and… consolidation, renewal and growth in keeping with *regional* identity and the aspiration of citizens" (Council for European Urbanism, 2003).

According to the zero km conception, another value that has to be changed is the importance of the *local over global*, which mainly denotes the preference for regional products and services. This can create a new application of the bioregionalist concept to new construction choices, that is, materials, techniques, and design strategies.

It can also be stated: "if in the medium-long period the globalization gives advantages to the economies themselves, in the short period it will encourage instability and uncertainty" (Mattoscio, 1999: 21).

b Materials for sustainable technologies

Within the approach to zero km design strategies, a fundamental role is played by the selection of materials, which are used for processing the sustainable technologies. Thus, besides and beyond the satisfaction of users' requirements, the architects, the engineers, and the planners, during the decision process in choosing the materials, should take into account the fundamental parameter of the local availability and its distance. Moreover, the level of *naturality* and the artificiality should be identified, according to the resources' limited availability.

The choice of material is fundamental because it can create immateriality, even out of concrete matter, thus providing spiritual values to architecture and to urban design. As stated by the Japanese poetic architect, Kakuzo Okakura: "only in the vacuum lays…the truly essential. The reality of a room, for instance, was to be found in the vacant space enclosed by the roof and the walls, not in the roof and walls themselves" (Okakura, 1906).

Even within the more concrete and massive works of some architects such as Louis Kahn, it is often possible to find a lot of "inversions: masses which suddenly seem weightless; materials which dissolve into immateriality; structures which reverse load and support, rays of light which reverse the realm of shadows; solids which turn out to be voids" (Curtis, 2012: 78). This is very clear in some of his works such as

> the Salk institute (La Jolla)…which uses an open space to address the horizon line of the Pacific, and which employs a channel of water and light to suggest a metaphysical dimension in the research into the hidden laws of *nature*…one recognizes Kahn's interest in the geometries of nature, including crystals and snowflakes…Kahn penetrated the substructures of the past and transformed them into the resonant emblem of modernity.
>
> (Curtis, 2012: 80)

> Besides the immaterial character derived from the space configuration, the American architect had always declared: "it is not [with] the stone and the wood, but *with light and air*, that I have marked my passing."
>
> (Curtis, 2012: 80)

More recently, the famous couple of alternative American architects, the Vales, who had first advanced the idea of building within the cycles of nature, without being slaves of fossil fuels, declared:

> This kind of architectural work came from a design and construction procedure in which the spaces studied and the material used were going far beyond a formal research aimed at itself...The comfort [is] an adaptation to acceptable conditions while remaining as close as possible to nature's rules...The Vales consider also the need for reducing at minimum the ecological footprint (of a construction)...during its whole lifetime.
>
> (Vale and Vale, 1975)

Electric energy is made up with photovoltaic, water from rain, and the vegetable garden with the wastes, the materials have to be at high naturality and/or recycled. It is suggested that, even if it

> is very difficult today to convince people to act, since there is a growing desire of comfort, of bigger houses, of ownership of more material goods...it is [nevertheless clear that it is] not possible for a *finite* planet to sustain a growing population with increasing demands of material goods...Our research had demonstrated that it is not possible to sustain the present life standards...A sustainable future is not possible without introducing meaningful changes in the behaviours [within their work, where the material choice plays an important role, as they] are trying to obtain the maximum with the minimum of resources. The research demonstrates...that not always the sophisticated technologies [the so called hard] lead to a better constructive return [but that] the use of a passive and not mechanical means for providing comfort to buildings appears more reliable.
>
> (De Lucchi and Pastenga, 2008: 4–9)

It is well known how, during previous decades, the consciousness of the damages provided to the environment by humans had created the need for identifying and applying a number of strategies, methodologies, and standard measures capable of reducing the energy and prime matter consumption by saving natural resources and by employing renewables. The engineers, the architects, and more generally, all the transformers of land had become more responsible towards the limitations of serious environmental impact, that is:

> The qualitative and quantitative, direct or indirect, at long or short term, permanent or temporary, single or cumulative, positive or negative, alteration of the environment, connoted as a relationship system between the anthropic, naturalistic, chemical, physical, landscaping, architectural, cultural, agricultural and economic, following the application on land, of plans, programmes or projects in the different stages of their completion, management and disposal, as well as of eventual misfunctioning.
>
> (Article 5 of the Italian Law Decree of 16 January, 2008, n. 4.)

In a few words, the impact is made up by "the whole of the alterations of environmental factors and systems, as well as of the natural resources, produced by the land use and human settlements transformation" (Bettini *et al.*, 1984: 19).

> The nature's economy is characterized by matter's, energy's and information's flows, then even our anthropic field should act in the same way, by means of sophisticated systems: the technologies; and stronger in nature the economy of richnesses results (there is no storage), higher within the human action the consciousness should result that no living system can exist which produces wastes not capable of being transformed naturally into goods; [in fact]…the welfare of one species cannot be at depletion of others; [and finally the] biosphere, in its neverending job of generating life, of producing order from disorder, of arranging things from the dispersion…does not consume anything, either matter or energy: it does not produce wastes. It does not destroy the energy received by the sun, does not consume it, but it irradiates it again into the space…so as all the matter which it uses is spread for employing it again…the cyclicity is another character of the nature's economy. [In few words] the nature's economy is guided by an ordination principle which emerges spontaneously in the ecosystems, [and the results of this method are clearly demonstrated by the fact that evolution brings more and more towards] situations of increasing collective advantage and of a more higher efficiency in the use of available resources.
>
> (Masullo, 2013: 30–5)

Within this referee frame, the design is located as a buffer in the relationship between human and environment that "has been established by an original bond… and in the denial of any casual link"; in fact, humans are not only part of the environment (as well as the environment is not only a datum), but at the same time, are "an un-objective and neverending changing dimension. [And yet] all this does not prevent at all to operate…on the contrary it allows to yield those modalities, which can be more suitable to the actual world complexity" (Ciribini and Gasparotti, 1990: 48).

When discussing materials in the construction process, the need arises of identifying, within the field of the materials' properties and of the products' performances, the biological qualities that limit environmental impacts, and the final ecological footprint provided by any transformation work, either as an architecture building or as an urban regeneration process. Whenever an architect, an industrial graphic designer, an engineer, or any other project maker interacts with the matter to transform it into use as art objects, buildings, squares, or gardens, the knowledge of the impacts that any operation and any process it involves is included in professional responsibilities. The knowledge terms, conditions and modalities of the materials' properties, of the products' performances, and of the technological units' behaviors, are studied within the frame of the *naturality* characters (see Abrami, 1987: 119–21).

The habit requires that the designer activated the use of materials with common sense as starting point. *Common sense* is not intended as a loss of *esprit*, on the contrary, it means, according to Voltaire's words:

> *Sensus communis*...not only but also humanity, sensitivity...but from where does the expression common sense come, if not from the senses? Men when invented this idiom were convinced that nothing entered in the soul if not by means of the senses; if it was not so, how should they have used the word sense for identifying the common reasoning?
>
> (Voltaire, in Segre, 1995: 288)

Thus, common sense can become a tool for the management of the design products, giving back to nature the role of defining knowledge. In fact, being the last active, perceivable project can make a work durable over time and useful (or harmful according to the kind and amount of impacts) for those who perceive it. In fact, the matter, in itself, is neither good nor bad, "it is as the clay under the potter's wheels" (Voltaire, in Segre, 1995: 288); it is the design-er's task to benefit from the work and to employ it at its best, by achieving the best goals and reducing the damages (the impacts). We can again affirm with our philosopher that not always the final aim is the destined one: "here are the stones which an architect did not make...," that, in some cases, the material choice can take a negative character: "the tools given by nature to us cannot always be the final cause in action...ears for hearing the sounds, the eyes for seeing the light," (Voltaire, in Segre, 1995: 156) while the designer's abilities and talent are seldom not enough, as "the eyes given for seeing are not always open" (Voltaire, in Segre, 1995: 157).

c Sustainable products

According to the principles that define ecosustainability and the biocompatibility of the construction, as well as of the design procedure, the criteria and perfor-mances that should be owned by a building product are to be clarified.

This question needs to consider material sustainability in general and, in particular, consideration of waste. Therefore, the principal supply of materials, that is, nature, should no longer be interpreted as primitive support spoiled by human's work and cancelled forever, but as the product of a long process of manipulation carried out by human's work upon time. Nature, in the meantime, has not been a passive and inert reality. It has interacted, accompanied, or con-trasted the transformations enacted, and has been, through a labor of an under-ground chemical laboratory with the decisive and free help of the sun and rain, an active protagonist of economic life. Only if this active role is recognized to nature can the economy be redelivered to its actual dimension, which centuries of economic theory had delayed and removed. In this effort of changing the nature, it can happen that humans, according to what Marx already reminded us, can find their self deeply modified. How can this new vision of people's actions of the transformation on earth affect the selection of materials during the design process?

The first operation is a responsible decision: the only way for achieving the consciousness, of course, is by analyzing, studying, and thus, getting a deep knowledge of the question of products' selection, both in strategic and practical aspects.

The entire product's life can be seen as a total of activities and processes, each of which absorbs a certain amount of matter, energy, and water, operating a number of transformations and releasing solid, liquid, or gas emissions which, once introduced in the environment, can cause pollution. It is due to the product's impact, that is, its damages and/or alterations inevitably provided by its lifecycle, that its weight on the environment and the following contribution to its ecological footprint are judged. When the alterations are balanced by the answers offered by the environment, and thus, are still included in the *carrying capacity* of that specific system, then it is possible to declare that those products' lifecycles are sustainable (see Wackernagel and Rees, 1996).

Within the frame of material knowledge, some technological innovations for building products can enlarge the body of possible employment for the human habitat construction to the use of ancient and forgotten prime matters. The latter, still very appropriate to the Mediterranean region, are for example the rammed earth or the bamboo, or some innovative products, such as the biocomposites or the organic resources for producing photovoltaics or micro-aeolic systems (see Carbonara and Strappa, 2013).

The products assembled with the previously mentioned matters now start to be appreciated by the designers for the number of advantages, not only in terms of sustainability, but also of bioregionalism and of their intrinsic and architectural qualities. For example, the *biocomposites*, emerging products aimed at being employed in various fields of the environmental transformation, stand as a valid alternative to the less ecological bi-component composites, and can be part of the green-economy market. They are made up with two or more distinguished components or phases, assembled together to shape a new material with very different properties from those of the single constituents and with a better performance (see Fowler et al., 2006: 1781–9).

Some traditional typologies of biocomposites have always been employed in architecture, such as the mixed systems in hemp and flour or the straw and earth, while other innovative typologies are increasingly developing, such as those at polymeric matrix. The latest are made up with a reinforced phase, which can include vegetable fibers such as cotton, linen, hemp, kenaf, recycled timber fibers, paper trashes, or even elements derived from food crops and fiber of regenerated cellulose. The matrix (the other component of the biocomposite) can be made of peculiar polymers, so-called biopolymers, ideally derived from renewable resources such as vegetable oils.

Another product that can be considered as sustainable, at high naturality, and coming from an ancient tradition is *bamboo*. It provides a good response to the modern design requirements as a valid alternative to reinforced concrete, and it does not create waste, as any of its parts can be employed for a different scope and can be recycled. This system, classified in the vegetable-materials class, presents many good advantages such as biocompatibility, ecosustainability, fire resistance, and resistance to seismic actions. With its tubular shape, it is suitable for a number of structural requirements, but it can also be employed as a reinforcement for concrete. Bamboo, employed for a long time in South America and South Asia, where the plants spontaneously grow, has recently arisen interest in other regions of the world due to its versatility. In Europe, where some plants have rooted well, diffidence is still high, even though—with the great lesson of the master Kengo Kuma—some designers have started to be less agnostic.

Another example of a highly sustainable material is derived from an ancient technique that used to exploit, in many regions of the world, the potential of the local earth for building walls, arches, partitions, and vaults. Generally known as *raw earth*, because in comparison with brick, it is not cooked in the oven, but it is left drying in the open air. It has recently become an important material for construction technologies because of its rediscovered properties, being both biocompatible and ecosustainable.

Some significance has been recently been attributed to construction of this material. In fact, even though for a number of centuries it has been established as a poor material in contrast to concrete (considered as the material for the rich), today, its qualities are being discovered by companies, constructors, and designers. This is due to the high capacity of adaptation to places and the conservation of local identity, besides the well-known properties of health, comfort, complete recyclability, and absence of environmental impacts.

Cultural reflection and human sensitivity lead towards new fields of applications, which are different from conventional ones. According to Heidegger's thoughts:

> The memory, *An-denken*, acquires for man a reassuring function, a possession sense, for the past. So even the technique triumph of the modern man, as well as the storicistic mentality triumph, becomes a re-assicuration of the position of the present (I am) in comparison with the nature and the historic past.
>
> (Heidegger, 2001)

An invasive technique could create a strong security in the human soul. In Nietzsche's thought, the security tool is located in the need for "re-constructing historically the roots, since [it is needed] for dominating the nature" and it can be recognized as "*power will*" (Nietzsche, 1924). The employment of the rammed earth could represent, on the contrary, a sign of the new consciousness of the need for abandoning the ambitions of power. For this power is no longer suitable to a world which the technique can control at any level, and in which there is no need to assault nature, but rather to safeguard it, in both its living and cultural shapes. A number of architects are rediscovering the rammed earth's technical and configurative qualities, besides those of image.

d Assessment methods

The selection of materials during the design process requires the employment of specific and scientific tools. In more recent years, such tools have been refined to verify and measure the various scales that the impacts provided by the anthropic transformations on humans and the earth. These are based on the studies of materials and technologies' behavior, as well as on those actions responsible for the various levels of pollution. The effects should be measured by means of scientific and statistic knowledge upon the various environmental sectors (water, heat, sound, light, air, soil, etc.) to establish a number of indicators[1] aimed at providing, in qualitative and quantitative terms, the state of the physical and chemical phenomena. The assessment methods usually employ these indicators and their units, which are able to represent the model of the building process, at the various scales from the single fabric to the urban areas.

The multicriteria system is the more common methodology for applying such methods, as it allows the management of a great number of data, both related to building and materials/technologies during the interface with the hazardous thresholds in various stages of the process.[2] An important strategy for assessing the impact is the ecological footprint (EF), the indicator created by Mathis Wackernagel and William Rees a number of years ago. It has now become the way of measuring people's impact on the earth (see Chapter 2, note 2).

This situation brings the countries of the world into a disparate situation regarding the actual fault of pollution on the planet. The evaluation of richness and importance of any country within the global assessment of countries' power and rightness to decide for the whole needs to be considered. Presently, the globally accepted criteria are opposite: the countries, considered the richest, are more impactful, that is, consuming more resources and producing more pollution, devastating the global reserves of pure air, clean water, and uncontaminated soil.

In order to reverse the situation, a number of economic actions have been developed, for example, the *polluting tax* that can push the rich, dirty, and invasive activities of the people's economy either to reduce their pollution and use of prime matters (energy, materials, water, soil, air, etc.) or to pay more taxes. Thus, contributing to the funding for depurating, purifying, sanitizing, and clearing the contamination.

Recently, Rees has also defined the concept of the "ecological debt [as] the domination of the rich persons over the resource sack, over the lack of biodiversity, the environmental damn, and the free occupation of the space for dropping wastes" (Wackernagel and Rees, 1996). According to the need for reducing the ecological footprint, important methods arose, such as the well-known American Leader in Energy and Environmental Design (LEED).[3] This system, particularly suitable for the urban design, was developed by the United State Green Building Council (USGBC) in March 2000, as a green building certification system. LEED provides building owners and operators with a framework for identifying and implementing practical and measurable green building design, construction, operations, and maintenance solutions, addressing the complete lifecycle of buildings: design, construction, and operation. Once obtained, the certification provides independent, third-party verification that a building, an urban area, or a settlement were designed and built using strategies aimed at achieving high performance in the five key areas of human and environmental health: sustainable site development, water savings, energy efficiency, materials selection, and indoor environmental quality.[4]

In another method, processed before the LEED, the GB tool, indicators play a variable role, by ending up in a multicriteria matrix. The GB tool is divided into two sections, the definitions and the software for the assessment. Its aims are that of "providing approximated assessment for a wide range of potential parameters for the environmental performances concerning the threshold values, for the region and the use destinations."[5] Both the quantitative entities and the qualitative indicators are represented in terms of scores, from minus two to five. According to all of the data and the information emerging from the building or the urban area knowledge, the scores are assigned and the model provides the resulted assessment for the level of impacts.

Another methodology can be mentioned, developed by a French team since the 1990s and guided by the local authority, the *Haute Qualité Environnementale* (HQE; high environmental quality). By means of a number of required elements, listed in the shape of recommendations and standards for the designers, three categories define the impacts, according to the scale (internal, local, regional, and global). The requirements (such as comfort, noise, water economy, ecological quality of the site, greenhouse effect, etc.) are linked to the actions to be made before, during, or after the erection of some built element, to be able to evaluate the possible future impact.

Finally, the *Valutazione Ambientale degli Edifici* (VAdE; building environmental assessment) is a model proposed in 2007, developed in order to deepen the effects upon the two targets (environment and human), and thus, divided into two large categories of ecosustainability and biocompatibility (Francese, 2007). Within the matrices of support, a score is assigned for each typology of impact, coming from the various actions provided during the entire building lifecycle. It is then possible to investigate nature's phenomena that interfaces with the anthropic processes, so that people "thanks to his presence in the field of carefulness," take care and bother. Therefore, generating "change towards the environment. And so it is the man, who is responsible for the change itself, who is the agent of any operation of environmental impact" and will not be "meeting-collision of – and between – things, but rather, through mutational provisions of human matrix, it will be meeting-collision of – and between – men" (Ciribini and Gasparotti, 1990: 49).

All these methods are applicable *ex-ante*, ongoing or after the design and/or the completion of the architectural or urban work, and are therefore, useful for simulating the future impacts' provision.

e Saving the soil

The role of soil as a resource for humans as well as for the biotic and abiotic world should be clarified when talking about urban regeneration, sustainable design strategies, and environmental technologies. The idea of the ecological footprint as a measure for human impact on earth is not casually taken with soil equivalent units. The use of soil, as an aspect of the ecological planning disciplines, plays a fundamental role within the chance of saving resources and avoids transporting them around.

Usually, soil is considered as "the sub-layer indispensable for the development of earth vegetation and as such it plays a literally fundamental role for the majority of the eco-systems" (Di Fidio, 1990). A more holistic sense of the word *soil* has been defined by the Centre for the Research of the Consumption of Soil[6] as:

> The open space, according to the modern planning theories, [which] represents the material for shattering the compactness of the historic city and for providing a new shape to its parts, to its settlements principles, to its building types; this is partly true, but with a cultural deep mistake: that of mixing it with a static vision of the land.
>
> (Vallerini, 2012: 56)

While the actual functions of this element are very well known:

> The soil provides support to other ecosystemic facilities, such as: production of food, biomass (wood, fibres, biofuels), habitat for biotic organisms, filter and regulator for water flow, container and degrader of polluting substances, contribution to microclimate, supporter for vegetation, aesthetic and landscaping functions, human waste and product's recycler.
>
> (Gardi *et al.*, 2013: 25)

If the soil is edified, thus impermeable, it no longer plays these functions; therefore, when planning specific changes into an urban area, which is usually poor in soil, mainly in the Mediterranean region, great care should be given in employing empty soil, and a comparison should be made with the social as well as environmental requirements.

The soil can thus be considered as *natural capital*, as it represents a vital resource, "which is limited and non-renewable, and on which the production of 95 percent of the human consumption is based" (Gardi *et al.*, 2013: 25). Attention has recently focused on this element, through literature and legislation, and it is considered a precious resource at the same level of the already acclaimed elements of energy and water. Reflecting on its potential, as well as on its performances and its hazardous question, soil can be defined according to the use made by human communities. If, in fact, within the urban questions, an *artificial area* is a surface used for residential, industrial, and commercial scopes, for healthy facilities, schools, community, for streets, railway, free time, and so on, the *soil consumption* involves any conversion of a nonartificial area into an artificial one. Conversely, if we talk about peri-urban (for peri-urban areas, see Part III) or country land, a *nonbuildable* area is part of soil employed for seeding, fruit trees, wood, humid areas, and so on.

It is well known that one of the main problems affecting the consumption of land is that when people build any kind of facility, building, or street, there is the need to change the soil qualities, both on the surface and below, to adapt its chemical and physical performances to the anthropic use destination. One of the main questions is that of water proofing, which can be considered as the permanent cover of a ground part and its soil with artificial material (asphalt or concrete) (see Munafò, 2013: 32). This treatment obviously creates the prevention of the natural water-cycle running, and as a consequence, dehydration of the deep layers of the earth surface, followed by an increase in evaporation in the air. The two phenomena provide an increase in temperature and a reduction of the clean-water tables' supply to the planet. In fact, it is clear by now that "we are still continuing to cover the earth's surface with concrete in a rhythm higher than 8 square metres a second" (Munafò, 2013: 32).

The question of soil can be considered on both the global and local scale. As far as the global questions are concerned, it can be seen that the desertification creates the impossibility of "supporting the biomass production due to the climatic variations and anthropic activities; the desertification is in fact generated by the progressive reduction of the superficial layer of the soil and of its producing capacity" (Vallerini, 2012: 55). Being well known that the desertification is due in part to climatic change and in part to the human misuse of soil, a number of measures have been taken by the UN against desertification. In fact, the

> UNCCD[7] establishes the need of national action plans in the industrialized countries, so as to create sustainable development with the aim of reducing the loss of productivity due to climate change and to land transformation for productive activities, urban growth and infrastructural development. [From these studies, the idea emerges that the desertification is] the decay, not only of the arid, semi-arid or sub-humid earths, but also of those which, even with various levels of depauperation, stand in urban and peri-urban zones, in areas dedicated to production activities, in fertile country areas.
>
> (Vallerini, 2012: 55)

Here, the local question arises strongly, and with that, the main study area, that is, the city. It is clear that

> the consumption of soil is associated to the growth of an urban area, often characterized by low inhabiting density and horizontal development…in areas which are often distant from the city centre and with…non-marginal presence of open spaces, which guarantee partial persistence of natural characters.
>
> (Munafò, 2013: 32)

It has been recently noticed how

> the surfaces of soil occupied and taken away from nature and from agricultural uses [were] enormously superior to the demographic growth and [were] governed only by the multinational market's requirements and by the land income...Facing up this scenario the sustainable architecture does not have any chance of determining any visible modification. [In some countries, such as Italy,] the comfortable land is devastated by urban outskirt...at the same time hundreds of small villages are de-populating...the youths left them in search of Good Volcanoes which erupt Nike and McDonald's. An enormous heritage, more or less natural and sustainable is left down, despite internet, the super roads and the big sheds built for pretending to obstruct an ineludible exodus.
>
> (Allen, 2008)

One of the last century's actions which increased the consumption of soil is the *zoning* strategy; the creation of the following is possible:

> The need for leaving the classical planning tool based on the *zoning* so as to take the innovative and holistic route based [instead] on the verification of the whole ecological operation and on the landscape-perceptive system, not only in the high-quality places, but also in all the territories with low value...and with no excellence and no restrictions.
>
> (Vallerini, 2012: 55)

However, even when the zoning strategy was no longer being applied, the intensification in the construction, and thus, the use of new and pure soil had still increased. In fact, it has been calculated that in Italy, from 2003 to 2008, 1,824,000 new houses have been built, facing a demographic growth equivalent to almost zero.[8]

Some strategies have lately been applied in order to face, and eventually solve, the question of soil consumption. One of the solutions is that of stopping "to construct, by withdrawing soil and starting to erect objects inside, over and with, the already built environment" (Allen, 2008: 2).

Even at European level, the 2020 packet contains strategies that take into account the use of soil, by reducing the consumption processes and soil decay. It is hoped that in the 2050 EU packet, a number of strategies would be included that would try to eliminate to zero the consumption processes and the soil decay.

The importance of soil in the regeneration processes of cities is fundamental, as saving space also means saving all the other supplies and resolves the social questions investigated here.

f The role of green spaces in urban areas

Associated with the question of soil consumption and fundamental as a solution for a number of polluting impacts due to several anthropic activities, the use of green spaces in urban areas deserves a complete book. However, at the same time, one cannot avoid mentioning this topic when talking about regeneration of the Mediterranean city with bioregionalist technologies and strategies.

Any city needs to be provided with a number of facilities and infrastructures, either directly or indirectly; it is nature that gives the prime resources. Thus, all the ecosystem facilities can supply the following needs. Feeding, by means of food, water, timber, and fiber. Regulation by means of climate, precipitations and air quality, and culture and recreation, such as social, spiritual, leisure, and aesthetic activities. Support by means of the water cycle, the nutrient cycle, the primary production, the photosynthesis, and the soil formation processes (see Millennium Ecosystem Assessment, in Peccol, 2013: 42–3). When the use of vegetation is widely promoted during the urban regeneration processes, besides increasing the most obvious and most important role, all the other resources can be exploited, including that of providing an oxygen supply to the atmosphere. The following describes the city management policy in Europe:

> The approach, based on green infrastructures, includes a network of natural and semi-natural green areas, open spaces, water course, punctual and linear elements of the landscape, in urban, peri-urban and rural areas, planned… with the strategic aim of valorising the multi-functionality and of achieving ecosystem facilities for the benefit of the environment as well as of the populations who live and work in a defined territory.
>
> (Millennium Ecosystem Assessment, in Peccol, 2013: 42–3)

This methodology promotes the use of green spaces as a producer of oxygen, but also the employment of vegetation for other uses, for example, green walls, green roofs, and green acoustic barriers. As it is known, trees play a great number of functions such as landscape, oxygen production, carbon dioxide retention, timber production, fruits, microclimate (wind barrier, humidity control, shading, etc.), slope control, etc. They can be classified into monumental trees, trees for timber production, and common essences. The timbers of excellence can be divided into long, medium, and short cycles of life, or in the very short cycle, the latest of which can be considered as timber biomass.[9] It is worth remembering how vegetation can purify water, soil, and air. However, it should be underlined that the use of vegetation at a building or urban scale can contribute to controlling the temperature, the sun, and the wind; therefore, regulating comfort and life quality, and reducing, by adsorption, the polluting substances in the air.

Even though vegetation plays a fundamental role in urban regeneration, which takes into account the enormous benefits for people and the environment, there is, nonetheless, the habit of assigning green spaces only to those spaces of the city that have remained free from other uses, and often only as the answer to the mandatory planning standards. A more sensitive approach to people's comfort would be to consider green spaces as a real design element, suitable for providing comfort for thermal, acoustic, and visual needs, and for enhancing the level of life quality in the city.

Urban areas are subjected to a number of polluting sources. For example, the intensity levels of noise—by interfacing with the auditory capacity of the target, usually the human—appears high in the city and comes from several different sources, such as the great concentration of traffic, presence of a market, hospitals, construction sites, arts and crafts, and small laboratories. It has been evaluated that the continuous, sporadic, and background noises coming from these sources affect urban areas up to 80 percent. The mitigation of the acoustic impact can be resolved through three strategies. The first is that of urban programming and planning aimed at the reduction or total breakdown of the primary noise sources, and thus, their causes. For example, an urban master plan that includes reduction of traffic, by avoiding or decreasing the concentration of the same activity in the same area, limits the continuous citizens' movement by car. The second strategy, applied when the first is not pursuable, acts directly on the target, in this case, the citizen and its vital space, by protecting it with insulating materials. The third and more employed strategy for sound protection is the intermediate, for example, the creation of barriers downgrading the sound wave propagation that started from the source, where it was not possible to eliminate them, and reached the target. The use of sporadic barriers only in the neighborhoods with strongest noises and employing plastic materials and technologies with dubious aspects, does not improve the urban quality, in either the image or substance. Instead, the design solutions of the sound absorbent floors or the vegetable barriers can reduce the sound intensity by three or four decibels. The best typology of vegetation suitable for noise absorption is that with large leaves, evergreen, with medium trunk, and with expanded foliage crown.[10]

The choice of employing vegetable barriers can satisfy other fundamental users' needs and health, as green spaces can also reduce pollution levels in the air. Parks and gardens can contribute to increasing the amount of oxygen in the air, as well as absorbing harmful gases. Barriers can also be employed in historical centers, wherever any free space is available. In particular, grass is considered a very good pollution absorbent, due to its capacity of attracting sodium silicate particles, strongly harmful to human health.[11]

Further employment of green spaces within the regeneration of a city is that of wind exploitation, both as a summer element of cooling and as ventilation control in winter. Moreover, wind can carry away sound and air pollution, as well as regulating the microclimatic conditions. With this viewpoint, vegetation can affect the wind propagation, as it can become both a barrier for air movements and a creator of wind phenomena.[12] In conclusion:

> The landscape and the open land, planned and managed by integrating with the urban vegetation, with the decay zones' recovering, with the private green, with the watercourses system, are *vital resources* and elements for re-launching the territorial, regional and local development and for favouring the environmental, social and economic sustainability, by reducing the ecological footprint of the urban areas.
>
> (Millennium Ecosystem Assessment, in Peccol, 2013: 42–3)

The selection of the vegetation systems for any urban regeneration should take into account the zero km approach, leading in the direction of autochthonous and bioregionalism essence, according to the previously mentioned needs.

g Selecting materials and products within the design decision for urban regeneration

The programs and projects for the regeneration of a city are a means for promoting and increasing the sustainability of the human habitat. Within this logic, the selection of products is one of the central questions, as it creates the chance of reducing the ecological footprint (with a limited use of new materials and new constructions), whilst helping to increase the comfort levels and the social sharing of public spaces.

The selection of materials, including the possibility of employing recycled products, can be interfaced with the role of policies and participating options during the decision process. They can be compared with the economic, environmental, and social factors that are to be found and evaluated within the city.

One of the main steps aimed at material selection is the acknowledgment that starts from the scientific analysis of the potential of the existing building materials. These potentials are due to the intrinsic properties of the composing prime matters and to the final performances of the whole product provided by the building market. Within this complex process towards the environmental consciousness, the designer's role includes a management procedure act to select, a little at a time, the system that appears more suitable for the future work's function and use.

In order to simplify the operations of products' selection and suitability to users' requirements,[13] the knowledge process includes the comparison between the various elements that shape the behavior of the materials and products and the number of parameters that define the project work (both existing or new). The size appears as one of the main requirements that the products, technologies, and any part of the process of transformation of the city should satisfy. This requirement, which was not restricted in ancient times, needs to be under control in Europe, where the resources are limited, mainly in terms of free soil.

This idea of restraining the size of any intervention has been introduced many times in architecture, planning, and ecology disciplines, and it perfectly matches the main environmental theories and methodologies.

For example, Jencks declared that "bigness…is one of the unavoidable characteristics of modern culture: the global market and ever-increasing populations seem to demand bigger and bigger buildings" but he argues, "bigness leads to boredom and anomie" (Jencks, 2002). Even Schumacher had introduced the idea of starting to reduce our aggression to earth with large elements (see Schumacher, 1973). To explain the importance of dimension in our culture, Jencks introduces the paradigm of the razor blade:

> The more razor blades are manufactured, the better each one is….but [he says that as far as the human land is concerned] there are limits to growth

and economics of size, a point reached when bigger means worse. The razor paradigm is not a model for all productive systems, especially artistic and complex social ones. When big becomes too big? [He invented the *Law of diminishing architecture*]; for any building type there is an upper limit to the number of people who can be served before the quality of the environment falls. [He declares that] money and size are measurable, dullness is not.

(Jencks, 2002)

After a certain measure (500,000 feet/46,000 square meters), the bigger the building, the smaller is the architecture. This phenomenon occurs exactly when there are inefficiencies in a system: the costs are multiplied by the dimensions. An extra cost for a floor (in a 40-storey skyscraper) increases the expense for each of the other floors as well. Usually, when a fabric is built for business, all the extras that are not strictly necessary are eliminated. These are the extras that define the things creating "the architectural art: a new concept, dynamics of space and light, ornament and structural expression, sculptural gestures and innovation" (Jencks, 2002).

Therefore, the sustainable designer, working towards either buildings or parts of a city, while giving attention to the size question, which is also responsible for the increase of the ecological footprint, should remember the following:

We need new geometries to restore human scale to primacy in architecture. As our ancestors who built the cathedrals knew – a tall building does not have to ignore the experience of body in space: nor does it have to generate the sense of anomie so common in the biggest development of the latest 100 years: perhaps the rise of sustainable technologies will help us to regain senses of scale and place.

(Davey, 2000)

Along with the size, another important issue of sustainable design is the depletion of the planet's reserves. It is well known how the reduction in consumption of energy, water, and the other resources in the construction sector creates an important role in the application of design strategies that take into account the resources employed during the whole lifecycle of the building, in the settlement, and the urban spaces. This care could improve constructive systems to obtain the highest energy saving, the best comfort for users, and the lower impact to the ecosystems by means of a sustainable process.

In an ecosustainable program, the selection of the building products should be extended to a time assessment. Besides the completion moment and operation time, the fabrics create effects both during the initial stage, for example, in the exploitation and labor of prime matters needed for the production of the technological elements, and during the following stages of demolition, disposal, and eventual recycle of materials and components.

Designing according to the cycled criterion can also favor the development of innovative materials, components, and building systems able to satisfy the sustainability expectations. A deep comprehension of building materials requires the designer to go through the complex recurrent process, which distinguishes the natural actions from the ones that have been applied until now by the anthropic civilization. This is assuming that "the planet provide resources for us, and the resources are transformed into waste, and nature has the extraordinary capacity… fed by the sun, to transform again the waste into resources." Then the production of objects, buildings, squares, and cities, should be processed from the concept of circularity, as it can be reminded that the "real challenge is today" that of finding the way of "satisfying the people's desire of continuing to have rich and wealthy lives, while at the same time being conscious of the fact that we have only one planet" (Wackernagel in Giddens, 2005: 93–4).

Notes

1   See "The Aalborg Chart," edited in 1994 and then under-signed by 1,860 local authorities, which defines the indicators as "administrative and city management tools for applying a sustainable model." It indicates the cities as areas "conscious of being responsible of basing their decisional and controlling activities, in particular regarding the environmental monitoring systems, the impacts' assessment, as well as those of the accounting, revision and information, over different kinds of indicators, including those of the urban environment, of the various urban flows, of the urban models, and the most important indicators of the urban sustainability." (Ambiente Italia (2003), *Indicatori comuni Europei: verso un Profilo di Sostenibilità Locale*, Ancora Arti Grafiche, Milan.)

2   For example, some of these methods are HQE, BREEAM, LEED. For the French method, High Environmental Quality (HQE), see the website, http://assohqe.org; for the English method, BREEAM, see the website, www.breeam.org; and for the United States method, LEED, see the websites, www.usgbc.org and www.gbcitalia.org (all accessed 31 October 2015). See Chapter 2, note 2.

3   See www.leeduser.com/credit-categories/3735 (accessed 31 October 2015).

4   The method can assess eight requirements classes: sustainable sites, water efficiency, energy and atmosphere, materials and resources, regional priority, indoor environmental quality, locations and linkages, and innovation in design. For each requirement, a number of requisites are listed that could be evaluated and to which various credits are attributed, according to the weight of the needs.

5   Integral translation by the author from the declaratory manifesto of the GB tool, whose complete title is assessment management system model.

6   See www.consumosuolo.org (accessed 31 October 2015).

7   United Nations Convention to Contrast Desertification.

8   Data extracted from the Italian agency CRESME (*Centro Ricerche Economiche Sociali di Mercato per l'Edilizia e il Territorio* - Centre for economic social marketing researches for building and land).

9   The tree with a long cycle is, for example, the baobab; the medium are the nut, olive, rosewood, and cherry trees; with the short cycle, the poplar; and with the very short cycle, the fir and willow trees.

10   The best plants for absorbing acoustic pollution are the *Cupressus sup.*, *Quercus ilex*, *Laurus Nobilis*, *Magnolia gandiflora*, *Carpinus betulus*, and *Arbutus unedo e Eleagnus ebbingei*, which present large leaves and a wide foliage crown. The tree or bush species can in fact decrease the noise levels thanks to a viscous and thermal dissipation created by the obstacle that the sound frequency finds on the leaves and trunks' surfaces. As the sound insulation is directly proportional to the leaf size and inversely to the distance from the source, the action of the barrier appears as more efficient the closer it is to the source.

11 Besides grass as an absorption element of harmful gases, some other vegetable barriers can be planted, such as the *Magnolia gandiflora*, the *Vitis silvestris* DC, the *Hedera helix*, and the *Laurus nobilis*. The air particles containing harmful substances, coming in contact to these essences, are captured and then left on the large leaves' surfaces. Afterwards, rain cleans up the surface, diluting the toxicity and again allowing the switch on of the cyclic phenomenon of adsorption and dilution. In particular, the most suitable shapes of leaves for absorbing the dust are the dissected one (Lamina), the simple leaf, and the compound leaf.

12 We can remember that the foliage crown is suitable for being a barrier for air movements and that, according to the latest's aspect, configuration and typology can define different pressure zones. The various shapes of a crown can protect from the wind. A horizontal developed crown defines a wider wind-shadow zone than a vertical one, while those with a round shape create reduced turbulence phenomena and friction. This deviates the airflow towards the upper, the lower, and the lateral zones; conversely, the conical developed foliage crowns can cause strong speed drops in the inferior areas and turbulences in the superior ones. Finally, the column typology of trees can create very limited wind-shadow areas. Not only the shape but also the crown size is an important factor for arboreal essences with thick foliage and they can provide notable variations proportional to the crown width and the height of the protected zone, while trees with reduced foliage do not determine differences. It has to be remembered that the protection from air masses can be guaranteed not only by means of tree-rows but also through a proper tree system, which can hinder the wind action, according to the reciprocal location. Another important issue is that of the summer and winter winds and the wind speed and frequency of their occurrence. This outlines the selection of more suitable trees, where it is required to create an obstacle in the airflow in the peculiar period of the year or to avoid its action within the whole 12-month cycle: in this latest case, it is necessary to employ evergreen trees.

13 The Italian standards UNI 8289 classify the requirements according to seven main needs: safety/security, comfort, usability, aspect, management, integration, and environmental safeguard.

4 Recycling and waste as a resource for urban and social transformations

a The question of urban waste

Provided that any manufactured object, building, or part of the city can present a specific level of naturality according to its lifecycle, properties, and to the number and quality of transformation actions, then it becomes clear how the products with high levels of naturality (and thus low levels of artificiality) can reduce the impact of extraction, consumption, emission, and conversion of gas, liquid, and solid substances occurred during their lifetime. Within this perspective, any wasted object, rubbish, or trash turns into a resource and the consequent actions of recycling, reusing, and composing again become the route to follow, apply, and carry on.

One of the main problems of contemporary city life is waste management, mainly in highly populated countries with a hot climate. A possible solution is already established as the opportunity to collect solid trash according to differentiated items and the reinclusion in the industrial and usage cycles. It has become clear that this question is strictly interfaced with a city transformation.

Attention should be paid to city life and management with regard to the concepts of reuse, the recovery of existing heritage, and recycling in the sense of reimmerging into the natural and anthropic cycles without producing waste. We have the support of the well-known American chemist, Paul Connett, who says that "the community has to say to industry: *if we can't reuse it, recycle it or compost it, you shouldn't be making it!*" (Connett, 2013). Connett argues that waste is a design problem. We need our best designers to draw up a way of reducing or removing waste from our system, as we are not able to perceive the value of the goods we throw away because we are dominated by the culture of excess, uselessness, and the ephemeral. However, this culture leads to a consequential increase in litter production.

Many experts and environmental scientists agree that we are living on this planet "as if we had another one to go" (Connett, 2013). While we should be conscious of the fact that we would need at least four planets if the whole world's population consumed like the average American and two if everyone consumed like the average European. Meanwhile, India and China are reproducing our massive consumption patterns. "If we want to move in a sustainable direction then something has to change" (Connett, 2013). In Connett's view, the best place to start that change is waste. Every day, each human being on this planet produces rubbish. A sustainable society has to be a zero-waste society. The zero-waste approach is better for the local economy (more jobs), better for our health (fewer toxins), better for our planet (more sustainable), and better for our children (more hope for the future) (see Connett, 2013). Connett has created guidelines for making this big change, which could also become a model to resolve the question of sustainable architecture and urban regeneration. In his solution, he outlined *Ten Steps to Zero Waste*, which are essentially common sense. He declares that a number of people would have little concern even dealing with the first seven steps, which are: source separation, door-to-door collection, composting, recycling, reuse and repair, pay-as-you-throw systems for residuals, and waste reduction initiatives at both the community and corporate level. "However, it is step eight where some people are going to have trouble and where, if we are not careful, the waste industry could easily co-opt all our good work."[1] The only possible route to follow in order to achieve Connett's dream future without waste is interpreting and imitating nature during the city's transformation.

> In other words waste is a design problem, and that is step nine: we need better design of both products and packaging if we are going to rid ourselves of the wretched *throwaway ethic* which has dominated both manufacture and our daily lives since WWII. We need to turn off the tap on disposable objects.
>
> (Connett, 2013)

Finally, one of the best solutions is step ten, which is the creation of "interim landfills…[in fact] the goal of zero waste initiatives is to eliminate the need for traditional landfills. These interim landfills should be seen as temporary holding facilities until we can better figure out how to recycle, reuse, or better dispose of these materials than just tossing them in the ground, and capping them" (Connett, 2013). Connett also integrates the three familiar *R*s of community responsibility: reduce, reuse, and recycle (including composting), with the less familiar *R* of industrial responsibility: redesign, which is the big doubt about future regeneration for a city. According to Connett's words:

> The first person that talked about zero waste was one of the greatest designers of all time: Leonardo da Vinci. Somewhere in his writing he said that there is no such thing as waste: one industry's waste should be another industry's starting material. No doubt he was copying nature's approach to materials. Nature makes no waste; she recycles everything. Waste is a human invention. Now we need to spend some effort to "de-invent" it.
>
> (Connett, 2014)

One can also add Antoni Gaudi as innovator and anticipator of the philosophy of reuse; he had employed in his works such as Park Guell, a number of secondhand tiles to make a masterpiece of the terrace balcony. There is no need to invent the idea of designing from recycled or reused objects, as this theory, as well as the following practice, can already be found in a number of experimentations and applications. The application of this approach to urban design, and furthermore, the employment of wastes coming from the city itself, is still to be completed.

Peculiar ideas for employing waste as a resource and for design have demonstrated not only that waste is available and reliable in terms of performance and use, but also that creativity, as well as the final image of the reused and redesigned object, is very often valuable in terms of artistic and qualitative properties. Therefore, it is desirable and easily accepted by citizens and users. A virtuous chain starts from differentiated disposal and continues with a number of new factories, specialized in this new manufacturing purpose, of redesigning, reproducing, and reinserting into the market these objects.

However, there is another way of recycling the natural and anthropic goods of the city. This involves the employment of existing parts of the city that have been neglected, abandoned, and forgotten, although it could instead be living again, if conveniently considered, studied, and finally recycled as a resource rather than thrown away as waste. In parts II and III, the description of case studies makes an effort to show how these examples consider waste as a resource even at urban level.

b The lifecycle assessment for city regeneration

In recent years, in order to allow any reuse or recycling of objects, buildings or parts of the city, the scientific application of the whole procedure has to be tested. This should be transparent, traceable, and clear, and should create limited environmental impacts along the whole lifecycle of the product, within both the product's life and the process chain.

Any study is based on the well-known system of the lifecycle assessment (LCA). It considers the whole product's life, from the extraction and acquisition of prime matters, through the stages of production and transformation of materials, water, and energy, until the stages of operation and end-of-life. Since the early 1990s, the International Organization for Standardization (ISO) has processed this methodology, which has been developed and evolved, until the LCA tools have become reliable and useful, and then converted into an international standard, codified by the ISO itself.[2]

One of the clearest definitions of the LCA system is provided by the Society of Environmental Toxicology and Chemistry (SETAC):

> An objective procedure for assessing the energy and environmental loads due to a process or an activity, carried out through the identification of the used energy and materials and of the wastes released in the environment. The assessment includes the whole lifecycle of the process or of the activity, involving the extraction and the treatment of prime matters, the manufacturing, the transport, the distribution, the use, the recycle and the final disposal.
>
> (SETAC, 1993)[3]

Being the model created for the control of the cycle of the production of goods and objects, it can also be useful for qualifying and quantifying the number of problems that any transformation process creates in the urban milieu.

c Reused objects

Although reused objects have been considered as waste for years, as they no longer played their task efficiently, they could now be employed, as they are, or partially recomposed for a different purpose.

If *to reuse* means "to use again or more than once," (from the *Oxford Illustrated Dictionary*, 1982) and to *reutilize* is defined as "to utilize a thing which has been already used, mainly for a scope different from the previous one" (translated from *Dizionario Garzanti Della Lingua Italiana*, 1981), then the action of reuse can be recognized in "any operation through which products or components, which are wastes, are employed again for the same aim for which they had been conceived" (Italian Decree, 2010. Dispositions for application of the European Directory 2009/98). Applying this easy concept to the production sphere of human objects, design objects, and built systems, the term reuse is stated as follows:

> [It] combines reusing materials and using items that have reusable qualities. Paper plates are an example of a non-reusable product. Cutlery that can be reused prevents waste at the landfill, but it also lowers the amount of energy needed to manufacture new products. Less pollution results and more natural resources are left intact. [The innovative concept is then to] consider the potentials of an item before discarding it, as it might be reused toward a different purpose than originally intended. An old shirt may become a car rag. Though reuse is different from reducing use, when an item is reused, consumption is reduced.
>
> (Peck, 2014)

In practice, the idea of reuse can be applied to several fields of daily life:

> For example, using cloth tote bags when you shop instead of asking for plastic bags, or buying reusable food containers, such as a thermal coffee mug or a reusable water bottle. It also means looking for ways to repurpose discarded items, especially those that cannot be recycled and will end up sitting in a landfill for centuries. Consider repairing an item rather than throwing it out. If you're upgrading an appliance or gadget, see if you can donate the old one to someone who can use it. If a product has served its purpose, look for alternative uses. For example, clean used jars can be used to store leftovers or odds and ends.
>
> (Wieman, 2014)

In particular, the question of nonorganic wastes within the city lifecycle could be faced by taking the opportunity of employing them as a useful second matter element for highly sustainable technologies during urban transformation and regeneration. This is without increasing the constructed volumes and the polluting effect of new materials and without exploiting new resources.

A number of studies and applications of wasted objects as reused products have demonstrated the potential of this idea. In 1975, when the Center for Maximum Potential Building Systems (CMPBS) was founded by Pliny Fisk, a number of new materials and products were developed, taking wastes as the base.[4] One of them is Ash Crete, an innovative building material made from the fly ash generated by coal-fired power plants and the bottom ash from aluminum smelters. The fly ash produced from burning pulverized coal in a coal-fired boiler is a fine-grained, powdery particulate material that is carried off in the flue gas and usually collected by means of electrostatic precipitators, bughouses, or mechanical collection devices such as cyclones.[5]

Following the invention of products manufactured from second-matter wastes, several applications have been carried out lately, mainly at architecture scale, but also at urban scale. Some examples are to be found in the Senegal Women's Centre (1996, architects Jenni Reuter, Helena Sandman, Saja Hollmén), in which many peculiar solutions were applied, originating from reused and recycled materials. For example, the frames are not made with timber, as the trees are a precious resource, but instead with recycled metals. The roof, tilted to favor the water release during the frequent tropical rains, is made up of a layered packet with a simple corrugated iron structure, placed on steel beams, with joints in recycled metal, and covered by an insulated layer, made up as a carpet in recycled straw, under the ventilated cavity, for cooling the underneath rooms. Some openings, made up with recycled tires, are in the lateral walls, while some windows are built with reused glass bottle bottoms (see Francese, 2007).

Cases that are more recent are the cardboard pavilion built by students in the Cambridge Fellows' Garden, Kings College, the Shigeru Bahn's paper house, or the lesser known, *Villa Welpeloo*. The latter, designed in 2009 by the young architects' office Architecten, founded in 1997, is located in the town of Enschede, and it has a structural system made up of 70 percent recycled steel that came from an old fabric factory. The insulation system is mainly obtained from a dismissed building, less than 1 km away; the timber is made up with old bobbins that came from a cable factory.[6]

It should be noted that the forthcoming habit of employing reused objects needs new standards and regulations. This will both create the consciousness of the route made up by the objects during their lifecycle (i.e., the amount of employed resources and the polluting effects), and the chance of gaining a certification, which will guarantee the traceability, ecosustainability, and health for users.

Within this frame of possible solutions for urban regeneration projects, made up with reused objects, several no-longer-useful materials or products can be identified, which regained the material culture's concepts of the Mediterranean tradition. These innovative technologies, based on partially reused objects, which come from urban activities, could reduce the hard content, either by consuming fewer fossil fuels during the relative lifecycle, or by employing those resources that are highly present in nature or also in the yield of secondary prime matter. Therefore, they will respond to the requirements of a high level of naturality, health, and the energy saving imperative.

d Recycling processes

According to current research and standards, the term recycle indicates products that derive from the same activity to which the new object is destined or from another category of activity. Different from the reused products, the definitions recall the need for considering recycling as the action of "converting waste into reusable material; of returning material into a previous stage of a cyclic process," where recycled products means "goods made out of recycled materials" (*Oxford Illustrated Dictionary*, 1982). Therefore, when a substance undergoes the same or another treatment or the same or another operation, it re-enters into the cycle more than once (see *Dizionario Garzanti Della Lingua Italiana*, 1981); thus, becoming more durable and a potential for saving new and precious resources.

The same legislation that established the definition for reuse also provides one for recycling, as "any operation of recovery through which the wastes are treated so as to obtain products, materials or substances to be employed for their original function or another" (Italian Decree, 2010. Dispositions for application of the European Directory 2009/98). The procedure of recycling, "recovery and reprocess of waste materials are applied for use in new products. The basic phases in recycling are the collection of waste materials [and] their processing or manufacture into new products."[7]

> The term *recycle* refers to the process in which an item or its components are used to create something new. Plastic bottles are recycled and made into carpet, pathways and benches. Glass and aluminium are other commonly recycled materials. Recycling is technically a form of reusing, but it refers more specifically to items that are discarded and broken down into their raw materials.
>
> (Peck, 2014)

The action of recycling also keeps usable materials out of landfill. Objects that might be considered waste are turned back into raw materials that can be used in the manufacture of other items. In particular, the following is known:

> Recycling consists of three basic steps. The first step is collection and processing. Communities handle this in various ways. For example, some may offer curbside recycling, while others may have central drop-off facilities. The recyclables are then sorted, cleaned and turned into marketable raw materials...for example, paper is turned into pulp and plastics are melted down. Step two consists of manufacturing the recycled materials into new products. Step three occurs when consumers purchase recycled products.

This completes the recycling loop. As a consumer, you can participate by properly recycling as much waste as possible and by looking for products that contain recycled content.

(Wieman, 2014)

The sustainable policy for waste management in the EU suggests prevention as the first option. As the best solution, aimed at avoiding pollution in the soil, water, and air environments, can be recognized in the absence of waste, a number of actions towards the reduction of the problem of wastes should take place, such as reuse, recycling, recovery of energy, and finally (as the least preferred option), the disposal into landfill. Nevertheless, more than 50 percent of waste produced in the EU is still disposed of in landfills.

Management usually includes the choice of reusing or recycling together with another two options. Landfill, in which nonperishable materials without toxic substances are stored after being thrown away. Incineration, which is the most practiced alternative to landfill, and recovers the heat released by combustion to produce electricity or combined heat and power (CHP).[8] Finally, recycling saves energy (generally less energy is required to manufacture products from recycled stock than from prime pure matter), and thus, emissions of greenhouse gases and other pollutants. It extends the reserves of finite resources (e.g., metal ores) and contributes to the sustainable use of resources. It avoids those impacts that are associated with the extraction of virgin stock.

Given this, by now diffused, knowledge on the subject, a great unexplored dimension of new potential scenarios could be easily opened facing waste management not as a problem to be solved but as an opportunity to be caught in a sustainable approach. In recent years, a number of initiatives were implemented by industries aimed at saving the growing costs of raw materials as well as at improving the efficiency of processes. One of the widely considered options has been recognized in the recycling of wastes produced as feedback during the same process.

Recycling materials from the municipal solid waste stream generally involves the following steps. Collecting the already differentiated materials from individual households and transporting them to a place for further treatment. Then sorting, balling, and bulking for onward transfer to reprocessors (e.g., at a materials recycling facility (MRF)). Finally, reprocessing them to produce marketable materials and products. A virtuous process for recycling usually includes the following actions:

1.   Identification of materials: the different kinds of materials present in any disused object are identified and then analyzed to evaluate their physical characters and chemical composition by laboratory tests.
2.   Projecting the process for material recycling: different treatments are tested to obtain new materials and new products, selected by a market analysis to evaluate the potential as a substitute for conventional materials.
3.   Tests: monitoring is carried out and a number of measurements are made to evaluate the opportunity to produce RDF[9] with those materials that are derived as wastes from the recycling process.

4. New project: a plan for both designing and processing a new system should be made for disassembling no-longer-used objects and for manufacturing new products.
5. Incentives for recycling processes: to program new cycled industrial processes for the whole chain of production; this facilitates, in the future, the chance of recycling and for exploiting the potential of the market and the waste management sector, also involving standard and law requirements.

The substitution of raw materials with recycled ones can save the embodied energy netted of the energy spent for extraction and conditioning of wastes.[10] Furthermore, the low cost of recycled materials can incentivize building insulation improvement, allowing a net reduction in the energy spent for conditioning. All these considerations imply a relevant reduction of carbon dioxide emissions.[11]

A further advantage of recycling is that of reducing the ecological footprint in building construction and urban regeneration. The substitution of new materials with recycled ones helps to reduce all the impacts in each step of the lifecycle. In order to actually apply the recycling procedures, it is known that various actions should be taken. These include sector analysis, aimed at reviewing and assessing the current condition and future prospects of the destination sectors for the use of transformed matter from disused objects. Then value chain analysis (VCA) is useful for accounting and presenting the value that is created in utilizing materials extracted from unused products as they are transformed from waste to raw inputs or components and to a final product consumed by end users. Finally, there is benchmarking analysis, with benchmarking of key indicators along the value chain to detect performance gaps and identify and prioritize constraints that directly affect the overall competitiveness of the value chain.[12]

As far as urban regeneration is concerned, the practice of employing wastes is still at its beginning, but already shows its potential, with the recycling products and materials either coming from other activities or derived from the construction and demolishing sector.

e The time issue

Time is considered as a fundamental question for the arrangement of the city asset, but also as a significant issue when the selection of products and materials for urban regeneration are considered for sustainability. Time affects the whole range of human transformation, which is a dynamic process, by setting the anthropic life in connection with natural cycles. Therefore, in order to restore these cycles in human habitat, any design solution can apply the sustainable development's idea that time and space are two faces of the possible future for the earth itself.

A number of reasons can support this idea, for example, that "time is necessary for the mind to observe and imagine and piece together the fragments that make up the space of life," which, reported within the architecture, can lead to an "ever-changing language of architecture [rolling] past connection between transportation, time and residential areas" (AAVV, 2013). In fact, if a worker saves time in going from house to work, then the building market improves and, among the other advantages, houses become less expensive. Other consequences are that if you waste time during your daily life, then work is less productive and if space moves fast when you travel, then the architecture can be watched and become part of the citizens' life and livability. For Baudelaire (the painter of modern life) modernity is "the ephemeral, the fugitive, the contingent, the half of art whose other half is the eternal and the immutable" (AAVV, 2013), and time is one of the most important issue of our age. In particular, in the architecture field, one of the crucial factors in the design procedure is time, as it is a tool for challenging the limits. As Aldo Van Eyck indicated: "whatever space and time means, place and occasion mean more. For space in our image is place, and time in our image is occasion." The transitory drawings, projects and the same architecture works are "displayed in time and space, becoming a culture of conversation whose indicators, depictions, fictions and tensions project the old into the present and the present into the future" (Mak, 2013).

Consider time as recycling potential. The idea of the avoidance of throwing away used objects and then reimmerging them into city life can be seen as a prolongation of time, normally given for the duration of any function to the object itself. The object's life stops to be considered as finite, and time stops to be considered as linear and short, but it develops its circular projections into the new functions defined by the recycled or reused waste. This means that not only urban regeneration should employ recycled and/or reused wastes, but also the process itself should be circular and temporary, by defining activities and events, conceived from their birth as transitory. Many an architect has also tested this idea of *temporary* during the whole history of human urban transformation, demonstrated by the interest in "an architecture of ideas, attitude and context rather than form, space and structure" (Jacob, 2009: 25).

f The transitional spaces

Given the role played by time within the city and its common areas, the regeneration process can test the production of temporary activities and transitory users. The interaction of appropriate technologies with transitional spaces could increase the quality of life in a city.

Promoting transitional spaces for European cities, and their streets, paths, and small squares, can satisfy the need for considering the human inhabitants of the earth as part of the environment and trying to reput the process of anthropological urbanization within the natural cycles, rather than employing the model of the straight line of using and throwing away without care for the future. The use of part of the present urban cycle (the waste from residential and commercial activities) as means of restoration, preservation, and maintenance of part of the past urban cycle (the historical centers and heritage) could become a new model for any urban development. This will abandon the idea of reconstruction *tout court*, without caring for the past, and instead, will create the chance of valorizing and using new parts of the urban heritage; this chance avoids them being left as they are, which in many cases means decaying.

The transformations can then be transient, mainly in the historical centers, where a number of limitations (legally in force) usually prevent any great change. Instead of leaving the decayed situation as it is, the transitional spaces could promote new activities, giving life to that part of the city again, and at the same time, creating the base for new development and social inclusion. Usually, these restrictions are considered as a means of protection as well as valorization of historical centers, in both their monumental parts and neglected areas.

The recent restrictions, nonetheless restrained as elements that characterize the historical city, just as much as the emerging monuments, can help to ensure the protection of the urban texture's identity. The transient solutions for designing these areas, which employ a number of byproducts, either certified or not yet classified, could provide a method of feedback for the definition of the appropriate sustainable strategies for the city. The transit proposals could apply specific strategies aimed at creating the need for development and requalification, as a nonpermanent system. The attributes of these strategies must match many requirements, such as the environmental, social, and economic sustainability; the respect for the historical context; people's needs; local potential; new users' attractiveness; reduction of systems to be constructed; maximum integration between the existing monuments and the historical texture; and maximum flexibility and ephemerality. The transitional spaces could then integrate the artificial and new elements within the context by carefully, and opportunely, selecting the technologies to adopt. The latter could, on one hand, achieve the improvement in the usage-potential of the area, and on the other hand, promote a careful reflection by designers on the materials to be selected.

The philosophy of the transitory architecture is supported by a number of already applied practices and studies, such as those of the *shadow city*, in which "tackling lifecycles of buildings and functions" can become "hybridized infrastructures of flows and organization…[where the designers] interconnected elements and systems transplanted from dissimilar contexts to suggest new propelling technologies." This will also promote social processors, which are "based on production lines, material cycles and flow patterns…experimented with flexible scaffolds and transient elements [so as] to produce surreal collision and calculated assemblages." In fact, "*shadow cities*, as kaleidoscopic centrifuges of urban histories and visions, [can test] how condenser could affect urban development, reprogramming and transforming over longer time spans" (Crary, 2013). This procedure of temporary architecture is especially appropriate when it is needed to regenerate "a zone which separates two entities, [which] often appears confused and with lack of organization" (Frediani, 2012), while it is known that the node is a vital crossroad.

> For not overcrowding such an area, a reversible project can be chosen, with *micro-transformations*, aimed at reconnecting the two parts of the area. Sometimes limits and barriers are there and cannot be removed, but can be seen in a different light; as opportunities.
>
> (Frediani, 2012: 12)

Several studies have confirmed that there are peculiar procedures to apply to these areas to avoid their decay and the reuse with a temporary approach is adopted at the same time.

Urban regeneration can be looked at according to the philosophy of a trilogy: reuse (which is similar to livability), reconnect (which is similar to connectivity), and recirculate (which is similar to walk ability). In particular, the character of livability can be defined as follows:

> An area [in which there are] not only physical and environmental, but also social, spatial and economic qualities…Achieving connectivity is important for success. Lack of continuity among the sub-districts of an area prevents them from serving as living spaces [and instead provides] negative impact on livability.
>
> (Yildiz, 2012: 26)

The most important quality of urban life, is the walk ability, which "tells about the pedestrian friendliness of an area." The pedestrian system should be "clear, comfortable, [with] direct access to all kinds of spaces," while the promotion of recirculation will establish "clear and defined routes for both vehicles and pedestrians [by separating] the two routes, [and then] avoiding confused interference between pedestrians, cars, buses, trams, coaches, metro stations, commercial lorries" (Yildiz 2012: 26).

A sustainable regeneration method requires analysis of the urban problems (visual and physical barriers, cars' access and parking, pedestrian's access, deterioration, and crime potential), and it is aimed at regenerating and redeveloping, by means of cultural and public buildings, which could catalyze the transformation. The historical background, where it is present, the location, eventual commercial activities, mixed use, views and vistas, and the waterfront, where existing, can all be taken as opportunities.

Another important issue for the regeneration of the European city is the fact that the entire historical center can be taken as an area to be recycled and reconverted into living zones. It has been already noticed how this will be possible only with transient solutions.

g Urban regeneration with soft technologies

According to the increasing importance that the time-and-space issue gets within the sustainable approach to urban regeneration, the need arises to clarify which technologies can be employed to reduce the ecological footprint as a whole, whilst at the same time, improving the quality of life. Such technologies, called at its earliest significance *soft*, are to be correlated with the need to reduce new constructions, integrated with the existing buildings and to employ materials at a high level of naturality. After industrialization and the resulting introduction in the building market of standardized products and complex technologies, the manual construction wisdom and the tradition of the material culture have been neglected.

This neglect had led to the materialization of some principles of architecture and so to the laceration between the fabrics, open spaces, and the external microclimate, also favored by artificial systems (see Banham, 1969).

A brief reminder of what technology means in architecture could help to dissipate all doubts and misunderstandings regarding the role of technology and the real contribution that the previously mentioned, so-called soft technologies, can provide to urban regeneration with a sustainable approach. In fact, if the technology is the art and study of using the technique, that is, the manners of transforming the matter into a product, or a useful object, by means of an anthropic tool, then there cannot be any harm in the use of technology, as it is only a means; therefore, neither good nor bad. This introduces the idea of considering the technology as a contribution to practical application of sustainable and de-growth strategies, if, and only if, the tools, methodologies, and products are selected among those considered *soft*, that is, friendly towards both the environment and the users. While examining these kinds of intervention the first concept to clarify is appropriation: to the place, to social groups, to climatic and natural milieu, and to economic and local potential.

The famous Gandhi sentence, "the technology should not create forms of exploitation of human beings," explains clearly how the link between people and the environment should be considered during sustainable procedures. The idea of appropriate technologies was born in the 1970s by transferring the concept from the cooperation works made in the (at that time so-called) developing countries. Including this concept into European city programs can clarify that, even in such ancient cultural background as Europe, it will be possible to re-establish the human and biotic life within levels of livability. The small-scale technologies, the cooperative systems, and the soft ways of procedures for production of goods and services can be applied in the Western city, not only in the small villages, with great advantage for the citizens' health and comfort.

> The appropriate technologies can resolve the requirements and the actual problems; they are strongly related to the local realm (social sustainability); they create easy reproduction with the resources which are available on site, and they are simply manageable (technical sustainability); they make a rational use of resources and reduce the environmental impact (environmental sustainability); they need low costs for plants and operation stages (economic sustainability).
>
> (Sorlini and Vaccari, 2009: 27)

The appropriate technologies should be included in a more general project of local regeneration and development, coordinated by a number of programs at various levels (national, international, and local).

In order to move in the direction of a completely upside down revolution in the lifestyle of cities and trying to achieve an actual sustainability, a progress can be made, from the aggressive, hard, and invasive technological means of the contemporary modalities of constructing, towards the appropriate systems, through the very well-known intermediate technologies.

> An intermediate technology will be highly more productive than the indigenous one and at the same time will be immensely more economic than the modern industry…The intermediate technology, moreover, should adapt to the environment in which operating. The machine should be simple enough and thus easy to use, so as the maintenance and the repair operations could be completed on site.
>
> (Schumacher, 1973)

Which kind of technologies can then be considered sustainable? It is possible to define the criteria for indicating the various categories of technologies, which, in accordance to what the bioregionalism suggests, to what de-growth requires and to what the local appropriation thinks, will promote a sustainable approach to urban regeneration. The criteria are mainly those considering the whole lifecycle process for the reduction of resource exploitation, pollution emissions, and health hazards. Some of these criteria can be:

1. Employing prime matter at high level of naturality.
2. Reusing, recovering and recycling as much as possible.
3. Reducing the amount of energy, water, pure matter, air, and soil, needed for the completion of building process.
4. Caring for the living beings of the planet.
5. Considering the global as well as the local questions.
6. Reducing the ecological footprint.
7. Integrating with urban questions.
8. Promoting social inclusion.
9. Limiting the fixed solutions at the utmost.
10. Defining possible future reuse and/or recycle of the employed solution.

Therefore, the sustainable technologies for urban regeneration can be synthetized into three categories:

1. The simplest technologies, with a high level of naturality, coming from the natural cycles and returning into them.
2. The attentive and deeply thought technologies that take into account the cultural background and try to be integrated with the anthropic process.
3. The solutions that compromise between the natural and the cultural environment, and which can represent at least an intermediate technology to be afterward substituted by a more sustainable one.

All the answers should be strictly in accordance with the use (and thus will help to reduce any waste due to the abundance of not needed resources) and be as flexible as possible. This will help for the future (and then sustainable) management, and avoid the need to make great changes and substitutions, which utilize new matter, fresh water, free soil, and pure energy (in particular the nonrenewable ones).

In addition to the previously mentioned essential requirements, the definition of products for new constructions should also allow the chance of extending the field of possible utilization of buildings for habitation. In addition, there should be employment of prime matters, ancient and forgotten, but very much appropriate to the Mediterranean region, such as the rammed earth or bamboo, as well as innovative technologies, such as biocomposites, organic photovoltaic, and microeolic turbines.

Finally, yet importantly, a strong parameter for selecting the technological solutions in urban regeneration is the character of transient nature. More scholars declared the need for nonconstruction and thus "to stop building [and to proceed] from the sustainable architecture towards the non-built architecture" (Allen, 2008: 2). The solutions are then to be designed as systems that, as a sort of infrastructural frame, will be capable of supporting those activities identified as compatible. At the same time, they should apply such technologies that should not be hard and fixed, should not be big and bulky, tall or built with very polluting elements, and should provide a low ecological footprint. They should employ as much vegetation as possible and green products, as a means of improving the quality of external air and eventually absorbing the toxic gas from traffic and other city activities, and they should promote the use of renewable energy sources. They should also be easily transportable and moveable, installed and managed, and perhaps be capable of being piled up and/or bent. Among the previously mentioned performances, another important aspect should be that of maintenance cost and ease, and thus, the technologies should be chosen from among the range of systems that will require little care and will show as little decay potential as possible. The recycling issue will have to be strongly pursued.

If, already in the 1970s, "the restricted notion of postmodernism…completely and uncritically reinserted architecture into the cycle of consumption" (Tschumi, 1994: 142), then it will now be imperative to stop considering architecture as a consumeristic sector and promoting the previously mentioned technologies.

h Social items in the recycling process

According to what has been said for the bioregionalism of people (see Part I, Chapter 2), the involvement of citizens within the processes of regeneration of cities is mandatory, and can achieve a sustainable and bioregionalist approach to the transformation aimed at providing any possible social benefits to the users. These benefits can be identified with the satisfaction of expectations, desires, and needs declared or not expressed by both the inhabitants and other people who utilize the area: citizens or immigrants from other districts or national and international tourist groups. The interaction between the various users' categories and the decisional authority is aimed at selecting strategies to be adopted for safeguarding and valorizing the urban areas under regeneration process. As it is known, the appropriate technologies for developing countries, but also any kind of sustainable approach, involves various environmental factors: social, economic, institutional, and technical. In particular, social factors can be recognized in the cultural level of the local population, in local habits, in the eventual disagreement between various population groups, and in the existence of local organizations or cooperatives.

One of the main steps of the process aimed at defining the appropriate technologies is that of the participation, obviously involving the citizens as part of the process itself. At this stage, the sustainability of the adopted solutions is measured in terms of social benefits and the achievement of social inclusion. If, in particular, the selected technologies employ reused or recycled wastes from urban life, then the social benefits can be recognized, for example, in the possibility of contributing to waste disposal and then quickly clearing and improving the local districts. An additional benefit is shown in the education role for younger generations in promoting the circular approach into the city.

Another social involvement can be found in the new idea of architecture as an event promoter and the role of space that the architecture itself will create for users:

> If the reading of architecture was to include the events that took place in it, it would be necessary to conceive modes of notating such activities... if movement notation proceeded from our desire to map the actual movement of bodies in spaces, it increasingly became a sign that did not necessarily refer to these movements but rather to the idea of movement – a form of notation that was there to recall that architecture was also about the movements of bodies in space, that their language (of the movements) and the language of walls were ultimately complementary ... architecture becomes the discourse of event as much as the discourse of the space.
>
> (Tschumi, 1994: 148)

i Urban waste as a resource for transformation

Finally, it can be declared with the words of Rita Levi-Montalcini (2009) that "the crisis of the modern society and its life and thought style, our production, wasting and consuming system are no longer compatible with people's and nature's rights." The urban transformation can then be considered as a chance and potential for interrelating with these modalities, by thinking and acting in a different way, starting with architecture as a space constructor. Space as possession taking, but also as dialogue, reciprocity, responsibility, care, and agreement.

The resolution of the duality between people and the environment has had various alternative occurrences in history. It consists of intervening with sensitivity and consciousness within the world circularity and in updating and keeping alive our knowledge of the deep links between people and the world, without excluding *a-priori*, in advance, all the components that we call *wastes*, only because humans are not able to make them enter into a circular logic again. Employing waste is a possible response to an urban quality of life that would also consider the need for reducing the growth of and balancing the *carrying capacity* of the earth.

The urban regeneration process is affected by a number of factors that interact with each other's and by political and technical decisions. If the role of waste is considered fundamental, the benefits can be evaluated only by comparing the waste policy with the factors themselves. The first category of factors, the citizens' requirements, which includes climatic comfort, the usability of the urban areas, and the health of spaces, can get benefits from the fact that the proposed solution can provide shadow, wind protection, and humidity control. Therefore, guaranteeing more comfort and happiness, with no additional constructed element to interfere with the natural milieu, the transitional spaces will help to favor, time by time, the better locations and site establishment of the recycled or reused products. Furthermore, the selection of controlled-chain products will guarantee the health and the absence of hazards.

The second category of factors, space requirements, which includes the maintenance need, the facilities' presence or lack thereof, mobility, and cleaning of public spaces, are to be satisfied by waste policy for a number of reasons. The first is that dirty things (the wastes) are to become a potential, then the recycled and reused objects can contribute to extending the tools of the facilities, without increasing the construction elements. Then improving the quality and the pedestrian capability of the regenerated areas can ease the mobility, and of course, the maintenance can be favored by the presence of activities on site and by the citizens' participation to city life. Moreover, all the place qualities, either positive or negative, can be valorized by the use of the recycled wastes, for these movable and flexible objects can safeguard the history and identity of places, as they do not involve changing the building and constructing elements, besides being part of the city itself and then belonging to its memory.

The space potential, as tourists' attractions, can be increased by the unnumbered objects that can be created by solid wastes, and that can generate useful parts of the tourist information, such as advertisement panels, signboards, benches, and resting areas. The traffic question, which interfaces with the chance of increasing the pedestrian areas, is helped by the use of transitory activities made up of waste recycled products, as these activities can recall people and help them to enjoy a part of the city that is not normally used. This is because it is often neglected and decayed, and thereby reducing the traffic concentration on the arterial roads. The aggregation spaces, which in the city usually establish a serious crux and an opportunity, can be valorized by the use of waste compounded systems, which will furbish and fill the empty and neglected squares and areas. The categories of bioregional characters of the city districts are identified with the morphologic situation, with a green presence, climatic milieu, acoustic environment, free soil, water supply, energy, and material resources. The waste-based solutions for the urban regeneration process can guarantee the protection of these characters for a number of reasons. The first of which is the fact that no additional resources are needed for satisfying the new activities' need. The second is that with the attention to the green areas, these flexible solutions are going to be bioclimatic, water saving, and renewable-energy users. The acoustic environment can be improved by the presence of additional objects, mainly where the laws did not allow building new things, but will permit the recycled flexible structures. The soil is going to remain free, if it was like that before, or can be covered by very light and transient objects.

Finally, the pollution question, crucial in the modern city, needs to be controlled because these new/old systems derived from the existing life will reduce the impact from the factories producing new materials and products. It will help to avoid on-site emissions into the atmosphere, soil, undersoil, and water.

It is clear that there is no new invention with regard to the idea of employing the wastes for city life, because at the margin of the urban areas there already exists a "juxtaposition of the *city*, as a flash-in-the pan contemporary terrain of anxious overproduction, controlled emergence and systematic diversity, and the *anti-city*, as a life palimpsest and a ghost space, a-systematic archive and junkyard of utopias" (Fedorchenko, 2013), where the habit of reusing and recycling has been in existence for some time.

It is now necessary to stop thinking that those who recycle and reuse should be considered poor, and start considering these people as the real environmentalists and future philosophers of this planet.

Notes

1   The incineration industry has discovered that by introducing two words it can continue to insert its poisonous, polluting activities into the mix. The phrase "zero waste to landfill" cynically takes the good intentions of the zero waste movement and moves it back in a non-sustainable direction. Instead, step eight calls on communities to build a residual separation and research facility in front of the landfill. The point of this step is to make the residual fraction very visible as opposed to landfills and incinerators that attempt to make the residuals disappear.

(Connett, 2013)

2 The standards that come from LCA studies and research are ISO 14040/2006 and ISO 14044/2006.
3 See www.setac.org (accessed 31 October 2015).
4 Pliny Fisk, by caring about the environment, has the mission to find new ways of reusing waste as a construction material. When he found out that the manufacture of cement produces 9 percent of carbon dioxide emissions globally, he wanted to find a new way of making cement. He wanted to stop the problems of carbon dioxide emissions by helping to reduce the air pollution caused when particulates or waste material are put into the air by coal-burning power plants. He decided to use these particulates from fossil-fuel-burning plants as the mixture to make concrete. Working in his earth lab on a farm outside of Austin, Texas, Mr. Fisk mixed the fly ash from a coal-burning power plant with some spoonfuls of water. He reported that the fly ash turned into a very hard substance in 20 minutes; during a more scientific test, the mixture demonstrated to be so strong that it broke his tester. Finally, Mr. Fisk came up with a concrete-like substance that is made up of 97 percent recycled material. This mixture is made from fly ash and bottom ash from coal-burning power plants, borate, and a chemical from the chlorine family (www.missmaggie.org and www.texasmonthly.com/story/pliny-fisk-iii-gail-vittori; accessed 31 October 2015).
5 In general, there are three types of coal-fired boiler furnaces used in the electric utility industry. They are referred to as dry-bottom boilers, wet-bottom boilers, and cyclone

furnaces. The most common type of coal-burning furnace is the dry-bottom furnace. When pulverized coal is combusted in a dry-ash, dry-bottom boiler, about 80 percent of all the ash leaves the furnace as fly ash entrained in the flue gas. When pulverized coal is combusted in a wet-bottom (or slag-tap) furnace, as much as 50 percent of the ash is retained in the furnace, with the other 50 percent entrained in the flue gas. In a cyclone furnace, where crushed coal is used as a fuel, 70 to 80 percent of the ash is retained as boiler slag and only 20 to 30 percent leaves the furnace as dry ash in the flue gas.

6 See the website of project at http://superuse-studios.com/index.php/2011/06/dag-van-de-architectuur-2011/ (accessed 31 October 2015).

7 Before any material can be recycled, it must be separated from the raw waste and sorted. Separation can be accomplished at the source of the waste or at a central processing facility. Source separation, also called curbside separation, is done by individual citizens who collect newspapers, bottles, cans, and garbage separately and place them at the curb for collection. Many communities allow 'commingling' of non-paper recyclables (glass, metal, and plastic). In either case, municipal collection of source-separated refuse is more expensive than ordinary refuse collection.

(Encyclopedia Britannica, 2015)

8 Energy recovered from waste can replace the need for electricity and/or heat from other sources. The net climate change impacts of incineration depend on the amount of released fossil-fuel carbon dioxide, both within the incinerator itself and during the process of fossil fuel consumed by the conventional energy sources displaced by the incineration. However, combustion produces emissions of carbon embedded into materials, thus, contributing to the greenhouse effect. Furthermore, incineration produces emissions of harmful airborne pollutants such as NOx, SO_2, HCl, fine particulates, and dioxins. Flying ash and the residues from air pollution control systems require stabilization and a particular disposal system is provided for hazardous waste.

9 RDF, or refuse derived fuel, uses a technology that produces energy from waste that is unsuitable for traditional recycling. Without this method of energy recovery, nonrecyclable refuse would simply be sent to landfill or incinerated with a negative impact on the environment. RDF captures the energy in nonrecyclable waste and turns it into a replacement for fossil fuels such as coal or oil. RDF or solid recovered fuel/specified recovered fuel (SRF) is a fuel produced by shredding and dehydrating solid waste (MSW) with a waste converter technology. RDF consists largely of combustible components of municipal waste such as plastics and biodegradable waste (www.iswa.org; accessed 31 October 2015).

10 In order to give a dimension of the achievable advantage, it can be considered that, for example, the embodied energy for the sheep wool is equal to 12.8 MJ/kg, that is, one-sixth less than that provided by the mineral wool commonly used for building insulation. A polystyrene solid foam has an embodied energy of 67.2 MJ/kg. By employing second-hand matters extracted from disused objects, a large amount of energy can be saved, whilst only the energy for conditioning will be used.

11 The EU report *"Waste Management Options and Climate Change"* (2001) evaluated that recycling 1 ton of synthetic materials allows a 41 kg of CO_2 eq emission reduction. Recycling 1 ton of textiles allows a 60 kg of CO_2 eq emission reduction. Recycling 1 ton of ferrous metal allows a 63 kg of CO_2 eq emission reduction. Incinerating, with energy recovery, 1 ton of synthetic materials can produce a net emission of 1,556 kg of CO_2 eq resulting from 2,257 kg by the burning process and 8 kg for transport, minus 703 kg for energy replaced.

12 As far as the waste management question in the local authorities is concerned, see the US Environmental Protection Agency (www.epa.gov; accessed 31 October 2015).

5 The contribution of an assorted expertise

How to guarantee subsidiarity and integration

a The various professions involved in the regeneration process

The sustainable regeneration process for a city requires absolute synergy between a number of experts who can interact and share ideas, methodologies, and the scientific capability necessary in order to achieve the common goals of satisfying environmental requirements. In addition, they need to provide integration and subsidiarity to the whole framework of citizens.

Architecture is not solely responsible for the damage to natural and cultural territories. Contribution to the whole process, starting with the large works of the nineteenth century until the present day, and innovation in material and product employment, could have followed a different approach and produced fewer environmental impacts by restricting and containing the ecological footprint values within a defined threshold. This objective could also have been reached by placing greater care on the scale of projects. The application of a sustainable development approach to urban regeneration today implies the use of new methodologies, mainly based on the cultural connection between the architecture, infrastructure, and city configuration, which can be achieved only by facing the interaction between landscape, planning, drawing, history, urban design, architecture, technology, engineering science, and building production.

Some different knowledge approaches are shown here, which are selected among technical or human sciences, usually involved in the regeneration process, before, during, and after the intervention. The relative actions are fundamental for achieving a sustainable strategy and bioregionalist care. Each of them can be integrated with the architect's feeling, meant at Voltaire's mode (see Chapter 2, paragraph d), and help them during the program: the drawing and representation are actions dictated by the sense of vision. Thermal and material science is more inclined to respond to the sense of touch, while the acoustic and seismic science responds to the sense of hearing. Technology and design composition satisfy the senses of touch, taste, vision, orientation, and vibration. Finally, the environmental sciences, as a whole—with the olfaction (the sense of smell)—help to guarantee the air quality, and pollution reduction. The environmental architect, the engineer of material, structure, environment, chemical, and hydraulics, the industrial designer, the sociologist, the planner, the waste expert, the energy experts, and the informatics specialists are all needed for exploiting the potential of a sustainable regeneration procedure.

b The role of social sciences: subsidiarity and integration

According to updated urban policies, both in central/northern European countries and in the Mediterranean regions, the city is defined as an engine of economic development. A space where most economic activities of production, exchange, and consumption take place. Therefore, any possible interchange of activities among citizens and other possible users should be promoted and safeguarded. During the regeneration processes, the first goal to be achieved, for the valorization of spaces, time and people, is that of guaranteeing both the subsidiarity and the integration among various groups of inhabitants. Consequently, the social aspects of regeneration should be borne in mind both during the investigation and analysis of the state of the art, and during the design stage aimed at proposing solutions and projects.

> The art of darning is the art of intervention in the *other city* which wants to re-sew the rip between the urban and the rural, the centre and the outskirts, the rich and the poor, the included and the excluded ones.
>
> (Persico, 2013: 11)

Any city transformation action should consider the social issues as parameters of the project and promote those strategies that try to achieve the following goals: urban redefinition aimed at social re-composition; technical and design quality of the proposal; sustainable building and energy requalification; and satisfaction of weak social categories' requirements.

Requalification, that is, the need for upgrading the spaces, places and the general urban texture to a level of good quality and livability for the citizens and tourists can be achieved by means of the following actions:

1. Not employing new resources, or, at least, restraining their amount.
2. Reducing the impact of demolition. (The building today represents the indoor space in which we live most of our lifetime and absorbs a great amount of energy and material resources, both in the productive and in the operational stage. Moreover, the garbage produced, both indirectly during the site development stage and directly due to the demolition, represents a quarter of all the waste caused by human actions.)
3. Saving the identity of sites that are historic, cultural, and part of the natural landscape.
4. Saving money, energy, water, and work.
5. Providing employment without increasing the surface of the built ground.

These actions seem to be necessary when the public open spaces are considered as the main part of the city itself, for they "aspired to [become] a *common good.*

The latest is still a source of verbal claims, and very often useful as a subject of discussion about the disappointment or abandonment syndrome, not able to promote experimenting actions of solidarity, cooperation and education" (AAVV, 1999: 16). Therefore, the social question becomes not only the main objective of an urban project, but also the actual engine for promoting a sustainable regeneration, where the architecture should be a mediator between the people and the place. However, sometimes

> regarding the architecture as an art at any effect could lead the designers, who practice it, to presume themselves as gods, as genial persons: a thing very often not true and origin of a lot of mystifications…What is missed in the so called contemporary architecture is the worry about what the users think.
> (Seminario, 2009: 42)

Therefore, the consideration of the social participation is a fundamental issue, rather than an option. "The bond to a community is founded on a whole of exclusions and inclusions, on the capacity of non-built walls between the various parts of the city, but membranes between dialoguing places" (Persico, 2013: 12).

Besides the usual and indispensable concern for social parts, the use destination, and the satisfaction of the requirements, the social subsidiarity requires the attention to be focused on the health and comfort of people. This can only be guaranteed with care for environmental protection. A number of benefits for citizens could be achieved following a sustainable regeneration methodology with transitory systems. For example, reduction of the environmental footprint, due to an increasing amount of recycled and/or reused urban waste will achieve greater collective relevance for local communities, in addition to the participation of the inhabitants in the whole process. There will also be cultural benefits, derived from the promotion and valorization of parts of town such as small squares, streets, and the paths of both the historical city and the outskirts.

The first step is investigating, describing and interpreting the social and cultural layout of any area in which a regenerative operation should begin, as the two aspects are strictly connected to each other.

In fact, "the culture is a right for all and it is a State duty to guarantee it" (article 9 of the Italian Constitution, see Buccaro *et al.*, 2014):

> It is a value that belongs to human history but fundamental in the present; being part of our identity, it is essential in order to produce welfare, to develop knowledge, to favour the social inclusion. The management of cultural goods is not equivalent to that of an industrial enterprise with lucrative goals, because its objective is not the financial richness but the cultural enhancement for a community: … and it is an essential factor of education to ideal, aesthetic and historic values, which they express as cultural goods: for its nature, each cultural place is distinguished according to the knowledge exchange in relationship with the surrounding territory and prolongs its proper action for the future generations. The culture is no person's, class', single country's

property: it is of all. Cultural Good means Common Good, part of a heritage, it is worldly.

(Buccaro et al., 2014)

The culture can then become a vector for programming with the right cultural policy. "The new point of the Italian Constitution tends to apply the principle of vertical subsidiarity, by decentralizing the valorisation interventions to local authorities, and that of the horizontal subsidiarity, by entrusting the management activity to the private entities" (Buccaro *et al.*, 2014). The social and cultural layout can be obtained by interviews, questionnaires, demographic profiles (by the agencies), sociological studies on the inhabitants and tourists, and finally, by defining the landscape perception of the city heritage.

c The drawing operations for knowledge and design

The first approach to urban architecture is conventionally made up by the tools of the survey, the drawing, and the representation. Therefore, the role of this expertise is fundamental, not only for the project, but also for the stage of knowledge, which, in the case of sustainable regeneration, characterizes the first set of data to be achieved. Often "the urban drawing anticipates the project, i.e. the system of intentions and relationships which had established the eventual completion of the project itself" (Raffestein, 2003: 31). It is also true that "architecture cannot divorce itself from drawing, no matter how impressive the technology gets. Drawings are not just end products: they are part of the thought process of architectural design. Drawings express the interaction of our minds, eyes and hands" (Graves, 2012). Moreover:

> Can the value of drawings be simply that of a collector's artefact or a pretty picture? No. I have a real purpose in making each drawing, either to remember something or to study something. Each one is part of a process and not an end in itself.

(Graves, 2012)

Therefore, the drawing role during the operation for requalifying the city is that of helping to process both the information and the ideas. Graves has argued that:

> [Architectural and urban] drawing can be divided into three types, which [he calls] the *referential sketch*, the *preparatory study* and the *definitive drawing*. The definitive drawing, the final and most developed of the three, is almost universally produced on the computer nowadays, and that is appropriate. What about the other two? What is their value in the creative process? What can they teach us? The referential sketch serves as a visual diary, a record of an architect's discovery. It can be as simple as a shorthand notation of a design concept or can describe details of a larger composition. It might not even be a drawing that relates to a [part of a city] or any time in history. It's not likely to represent

a *reality*, but rather to capture an idea. These sketches are thus inherently fragmentary and selective. When I draw something, I remember it. The drawing is a reminder of the idea that caused me to record it in the first place. That visceral connection, that thought process, cannot be replicated by a computer. The second type of drawing, the preparatory study, is typically part of a progression of drawings that elaborate a design. Like the referential sketch, it may not reflect a linear process. (I find computer-aided design much more linear.) I personally like to draw on translucent yellow tracing paper, which allows me to layer one drawing on top of another, building on what I've drawn before and, again, creating a personal, emotional connection with the work. With both of these types of drawings, there is a certain joy in their creation, which comes from the interaction between the mind and the hand. Our physical and mental interactions with drawings are formative acts. In a handmade drawing, whether on an electronic tablet or on paper, there are intonations, traces of intentions and speculation. This is not unlike the way a musician might intone a note or how a riff in jazz would be understood subliminally and put a smile on your face.

(Graves, 2012)

Furthermore, as Graves works with his

computer-savvy students and staff today, [he notices] that something is lost when they draw only on the computer. It is analogous to hearing the words of a novel read aloud, when reading them on paper allows us to daydream a little, to make associations beyond the literal sentences on the page. Similarly, drawing by hand stimulates the imagination and allows us to speculate about ideas, a good sign that we're truly alive.

(Graves, 2012)

The urban-architecture drawing or, as it is more suitable to the present role, its representation, usually can provide different meanings according to transformations due to the survey codes' modification and to the following interpretations produced by each cultural field. The urban sign, mainly for cultural conventions, is linked to definite meanings. The drawing's role is that of attributing these meanings. The sense of representation can then be enriched, renewed, or re-evaluated when different meanings (due to successive interpretations) are provided by a great number of people within a wide historic-geographic realm. Therefore, the sign can transmit even richer and more complex meanings than the original ones (see de Rubertis, 2008: 74–5). "In the prefiguration [of an urban space] casual attributions are going to be sediment of meanings, which the aesthetic and social conceptions in any historic age have given to the peculiar iconographic model used for the representation itself" (de Rubertis, 2008: 75). In addition: "the transmission of ideas for images is one of the fundamental bases of human thought and, if a visual organ did not exist…the concept itself of an image projected on a surface could not have place" (de Rubertis, 2008: 81).

The roles of drawing and representation within urban regeneration can be identified with two main actions. The first being the interpretation of the urban phenomena, due both to the sociocultural age and to the survey team's conceptual disposition that provides an image of the dynamic process. This could help to understand the existing peculiarities, the morphological identity, and the space value of the present situation of the place, while providing some local potential for transformation. The second action is previewing a future change with a project for regenerating the spaces. This role is fundamental because it is through the action of drawing, either by hand or by other means (various tools and/or computer), that the project can actually take shape and the scenarios of possible future changes can become perceivable.

Drawing is not only the image of the project; it is the manner of designing, thinking, and concealing the urban architecture itself. "The ultimate goal of the architecture consists only in making visible, to the intuition, now this, now that side, in symbolizing them and making them become representable through a human work" (Hegel, 1972).

> By denying then to [urban-architecture] representation a commonplace objective neutrality, in order to move its role from an aseptic and static tool towards a dynamic means for knowledge and communication, it is necessary that [the representation] can carry and include in it all those specific values which are identified as a sign, suitable for communicating the meanings of the space itself.
>
> (Sgrosso, 1979: 17)

The signs in the city are the basic elements for distinction and quality evaluation, so the drawing tools, which are aimed at identifying such signs, are essential during the learning process. Unfortunately:

> Very often, the drawing, as a representative tool during the analysis of the urban realm, is edged as a mere informative role about the data of metrical and formal nature, while a long series of essentially perceptive components are excluded and neglected. Components which are not only perceptive but also able to provide information on the space configuration, on their global image, on the usability and on the livability of the linked articulations.
>
> (Sgrosso, 1984: 7)

Drawing and the geometrical representation of the urban space has an authentic meaning of expressive language, and can be flexible and interpretative, and thus, be able to create a substrate of metaphoric or physical signs. The latter could become the base for the project, which is the interpretation of possible changes already incorporated in the reality. This is mainly true when the final aim of the project is a bioregionalist regeneration, where the existing meanings are fundamental and the drawing can help to clarify them.

d The history contribution for knowledge and design

Historical centers are seen as the social, economic, and cultural core of the European city question, as long as they are harmonized with modern lifestyles and information facilities. This usually requires a number of transformations that could compromise the identity of places and the historic value of the architecture. Thus, protection agencies all over Europe, aware of the problem, created a number of restrictions. Nevertheless, the regeneration process is often necessary, mainly in certain decayed areas, sometimes located precisely in the historical centers and sustainable actions should be interfaced with heritage existence and value.

> Once the heritage aims and those of the sustainable-habitat production diverge, an undeniable need is remarked of integrating them within a common system of analysis and reflection in order to approach – with the hope of success – the issue of the transformation of the contemporary urban frames which often contain some heritage elements already…In the organization of urban – past and future – developments, a number of variables are needed so as to join the issues of the heritage and those of the sustainability in the same project. These two dimensions, the cultural heritage and sustainable development, are often needed as reference at the detailed design scale, but also as thought for the global aims of a territory.
>
> (Carabelli *et al.*, 2011: 6)

In the current debate on the *historical city* (a concept that has now replaced the reductive one of *ancient city*), it becomes very important to recover any spread mark that still characterizes the urban textures, as signs of their structure and, even more, of their physicality. Latterly, these networking systems, created by paths and streets, often neglected as minor elements, have finally caught the interest of local authorities' programs and of the scientific experts' research, because they give identity to the city itself. The identity of any European city is very difficult to define:

> For recognizing the typical characters of the urban texture, by reading in between the lines of the building waves, and so defining within them the subsistence of significant traces through an analysis of the available documentary, iconographic and cartographic repertory [it is necessary] to identify the historic realm of the city… this is at the root of the conservation of the *memoria formae urbis* (the urban shape memory) and of a concrete safeguard and valorisation action of its cultural witnesses.
>
> (Buccaro *et al.*, 2014)

Nonetheless, some transformation actions can create a number of hazards in the city, which will cause the progressive and consistent loss of recognition of the urban shape and memory, with disastrous reductions in the social and cultural realms. From these hazards, the need for regeneration, rather than reconstruction, arises strong and urgent.

The role of the history of architecture during regeneration is two-fold: the contribution to knowledge, in the stage of the analysis, and that of the decisional stage of the project. During the analysis, the historical experts preview the studies on land configuration "since its origins, starting with the identification of the historic meaningful areas, are still provided with a proper physiognomy and recognition: so besides the very core of the city, also the outskirts should be considered" (Buccaro *et al.*, 2014).

As far as the investigations of the cultural milieu are concerned, it can be said that these data are important because the built parts of a landscape are strongly responsible for its identity, together with the natural components. It is necessary to distinguish between the existing constructions that will be key in the analytical evaluation—from the viewpoint of the functions, materials, the links with other elements, and the performance characters—and the historic and artistic emergences. These constructions are identified in urban sites such as squares, paths, districts, blocks, and fabrics' typology. Other processes of cultural origin should not only be defined but also carefully analyzed and, if possible, quantified, because, being of human manufacture, they already represent damage to the environment and to users' health. They include, for example, mobility, physical decay, functional decay, the downgrading or the lack of facilities, and inhabitant congestion. The cities represent fertile ground for studying the cultural phenomena and processes activated by human occupations, because we can easily find cultural resources, such as historic and high quality architecture, monuments, and other works that characterize its identity. They can be remarkably decayed by secondary polluting processes.

Conversely, the contribution of the history of architecture to the project helps during the policy of valorization and safeguarding of the cultural heritage. Therefore, pushing towards "the economic re-launch of the city…thanks to the, sometimes, unsuspected permanence of the ancient urban shape and of its historic, artistic and anthropologic peculiarities" (Buccaro *et al.*, 2014).

The latest innovative policies for safeguarding cultural goods have produced models of valorization of the cultural heritage with the shape of network organization and stable relationships that produce advantages, in terms of economy and arrangement, such as the complex systems, museums, and cultural districts. The cooperation between various local and national authorities can create attractive spaces to invest, work, and live, by promoting recovery and cultural valorization. In addition some benefits can be derived from these cultural resources, whose usability was not ensured, by qualifying and adding value so as to obtain positive impacts on the whole local economy (bioregionalism) and in particular on sustainable tourism.

A number of guidelines can be suggested for improving cultural goods management, such as the definition of the historic identity of the city through the analysis of documentary and cartographic fonts. In addition, there is the adoption of appropriate legislative tools for landscape restrictions, selection of opportune strategic choices for the *cultural goods* management, integrating the intervention in the territory, and finally, the adoption of a financial policy aimed at the valorization of the environmental, landscape, and touristic resources (see Buccaro *et al.*, 2014).

An innovative and interesting tool for analyzing the city background in all its parts is the so called microhistory, which can be defined as "the intensive historic investigation of a well-defined smaller unit of research (most often a single event, the community of a village, a family or a person)" (Iaconesi and Persico, 2012: 14–29). The microhistory, though, is not less relevant than history and it is not an effort to focus on smaller things, which can be framed in a wider context so as to be *significant*.

> History is, of course, the result of the progression of large-scale transforma-
> tions to the structures of human societies, their relationships, their arguments
> and agreements. But these large changes do not happen in a vacuum. They
> happen within human societies, which are made by human beings, and by
> their relationships, cultures, imaginations, desires and expectations.
>
> (De Certeau, 1984)

The role of the history of architecture in urban regeneration is evident and nonde-niable, as any sign of the past should be investigated, connected with the modern as well as contemporary elements and its future development should be previewed and controlled to avoid any incoherent action and/or aggressive intervention.

e Planning and landscape research

According to the new concept of *cityscape*, the cultural heritage of the built part of the city is interpreted as a place in which perception of and use by citizens is considered as a means for reducing the "never-ceasing development of construc-tions" (Ingersoll, 2003: 41) and other very polluting activities. In order to face the question of cityscape, an introduction is necessary about both the landscape and urban topics. The distinction of meaning between the word landscape and the usual other two terms, which indicate the space of human habitat, environ-ment and territory, can help us to understand how man's activities behave with geographical elements. In addition, how to interpret the interactions between the physical and tangible aspects (natural and artificial) and the psychological and spiritual ones that configure our presence on the planet.

If the *environment* is the "surroundings, surrounding objects, regions, condi-tions or influences" (*Oxford Illustrated Dictionary*, 1982). Alternatively, it is "the space that surrounds a thing or a person and in which this moves or lives, so identifying our boundary as a proper autonomous entity which guarantees the human beings' company during their life on earth" (translated from *Enciclopedia Treccani*). Then the study of the various components of the environment is the primary operation to know our habitat, to occupy it, love it, and thus, transform it correctly and with respect. The biological definition supports this hypothesis: "the whole of physical and chemical (temperature, light, presence of salts in the water and in the soil…) and biological (presence of other living organisms) conditions, in which the living beings can run their life" (translated from *Enciclopedia Treccani*). This idea can provide the chance for the natural and artificial elements to interact with the biological organisms. The figurative meaning of the environment, as "the whole of social, cultural and moral conditions in which a person stands and develops his personality, or more generally, lives" (translated from *Enciclopedia Treccani*), leads to the involvement of the social and cultural components within dynamic processes.

The term *land*, or the more inclusive term, *territory*, can attain a more concrete significance. Territory is a "land over jurisdiction of sovereign, city, state, etc., large tract of land, region and portion of country not yet admitted to full rights of a state or province" (*Oxford Illustrated Dictionary*, 1982). Land is a "solid part of earth's surface; ground, soil, expanse of country; country, state, landed property" (*Oxford Illustrated Dictionary*, 1982). In a wider context:

> [Territory is] a geographic region or zone, a portion of ground or soil with a certain extension: a big territory; a stripe of territory; the mountain territories; the coastal territories. In particular, extension of a country included within the administrative State frontiers or which sets as a legal entity; the territory of a state, in which the State itself practices its sovereignty, including, besides the dry land delimited by boundaries, also the under-soil, the internal waters, the territorial sea and its bottom, and the atmospheric space dominating both the land and the sea.
>
> (Translated from *Enciclopedia Treccani*)

This can lead towards an aspect limited to the soil characteristics (thus, the part regarding the hydrosphere and the lithosphere), without providing complex relationships with the dynamic processes and interactions, as it represents the state of the surface of the earth at a certain moment in time. From the point of view of the transformations and design works, which define its criteria and strategies, this territorial component plays a fundamental role, as it refers to a tangible presence, legally represented and physically concrete, which should be taken into account under different viewpoints: environmental, legislative and administrative, physical, human (and thus social), and finally, dimensional.

The last word, which differentiates substantially from the previous two, *landscape*, had assumed wide and various connotations, all strictly linked to the concept of perception. An old meaning is that of a "picture…or part of one representing inland scenery; actual piece of such scenery" (*Oxford Illustrated Dictionary*, 1982), mainly underlined on the sight, and identified with "sightseeing, panorama; part of territory which can be embraced with the glance from a determined point" (translated from *Enciclopedia Treccani*). This stresses the attention, even for the geographic disciplines, on the very sense of vision:

> The whole of elements which constitute the physiognomic features of a certain part of the earthen surface; it can be considered as the abstract synthesis

of the visible landscape, for it focuses them only on the characters that present the more frequent repetitions over a more or less big space, superior in any case to that included in a unique horizon.

(Translated from *Enciclopedia Treccani*)

Instead, the more approved and modern definition of landscape embraces all the sensorial, psychological, and spiritual components, linked to the landscape and territory's fruition from people. In the European Landscape Convention, in *article 1* of the official version, it reads, "*landscape* means an area, as perceived by people, whose character is the result of the action and interaction of natural and/or human factors." The importance of landscape quality in human habitat is underlined by the following definition:

> The landscape is a palimpsest which saves in the present and the future, the signs of past; territory is a parchment constantly cancelled and re-written, but it leaves, without will and without knowledge, some traces [of the past territories] of which the archaeology makes its dairy.
>
> (Raffestein, 2013: 33)

The cultural landscape includes, as it is known, both the natural and the anthropic components and it is configured as a harmonic system between the biotic (flora and fauna) and the abiotic elements of the territory (mountain, plan, morphology, soil, rivers, lakes, and climatic conditions). The modalities with which the studies and the following design for these landscape units are carried on requires an even more sustainable approach than the usual one employed for the actions towards situations and territories that are recognized as less fragile.

The fundamental role of landscape studies is that the landscape itself can be considered as a redemption, for the reasons taken up by Ingersoll:

> The conscience of an incumbent ecological hazard…[thus] any gesture which re-integrates natural environments and which participates to a conscious gardening…indulgence, a small step towards the redemption; a never-ceasing development of constructions…the residential zones left down to decay and to wastes…had often eliminated the contemplative experience of landscape from the daily life. While a big amount of empty space remains…and that sense of emptiness, which we can find in the beauty of natural sites is disappeared, since those holes are never framed, designed, conceptually distinct.
>
> (Ingersoll, 2003: 41)

In fact, there is a large consensus on the importance that the outskirts can play in the urban landscape, where the "transformation of rural areas in an urbanized territory should be defined as a landscape of dispersion, or as a *sprawl-scape*" (Ingersoll, 2003: 42). Therefore, urban policy is fundamental for safeguarding the urban landscape. Both in the historic central areas and on the outskirts, the urban planning's role within city regeneration is obvious and documented by the fact that "the urban planning choices regards the *common goods* such as the collective spaces of the city, the social facilities, the natural beauties and they affect the eco-system" (Moccia, 2012: 13).

In order to safeguard these common goods there are some conditions, without which it is not possible to make a plan today. It is necessary to be conscious of the involvement of various communities within the definition of *common goods*, and in particular of the collective space. In addition, strong differentiations among the communities should be taken into account, even not necessarily coming from various ethnic groups, but defining very different expectations and desires. Furthermore, communication systems and public opinion should be kept in mind, as well as applying decisional processes with the shape of dialogue and participation. The political practice of urban planning decisions depends on the modalities of the carried out studies, applied solutions, and the maximum times occurred (see Moccia, 2012: 14–15). When referring to the urban outskirts, it is important to mention the updated concept of the third landscape, coined by Gilles Clément, which deals with the abandoned and "interstitial fragment of the planetary garden." Furthermore:

> [It] indicates all the places in which man leaves to the sole nature the landscape evolution, where all human activities are suspended. [It] is everything which is neither light nor shadow, a residual, distinguished from both the spaces never undergone to exploitation (the primary systems) and those protected by the human activities (the sanctuaries). It is a fragmentary territory, charged with great symbolic value and still, nevertheless, residual, undecided, suspended. This space, as a wilderness, makes clear the breaks within the logics of appropriation, inclusion, specialization and taking advantage of, and/or capitalizing, the space. A refuge for natural diversities (to which we used to give the same, if not major, relevance as the social diversities which in the present urban milieu still show some distress).
>
> (Clément, 2004)

In the small abandoned historical towns of Italy:

> It is more understandable why Italy has become the unhappy ones' den. Nobody really does what he would like to do and in the place he would like to be. The life, once pastoral and agricultural, with the mule's breath between the small roads and the precipice had left way to asphalt and concrete. The shame of being ancient has been disguised with cars and apartment buildings. But if you undress yourself from the past, and remain naked, you cannot think of dressing yourself again with two buttons and a hat. It is necessary to find a texture and to spin it with patience.
>
> (Arminio, 2011: 23)

The role of urban and landscape studies can be synthesized as the participation of territorial questions into the regeneration of the city even when the portion of the interested land is small. This is because what really matters is the idea of involvement of the users and citizens in the procedure, as it is their presence that creates a cityscape.

f The design process

The project is an ethical moment of the regeneration process and involves a great number of issues. The bioregionalist and sustainable approach to the regeneration of a city requires a peculiar kind of design procedure that will listen to the involved stakeholders. It will take into account both the environment and the citizens as a recurring phrase of regeneration actions.

> To re-write the city's and land's history became a new narration, in which the diversity of the other cities' fictions within the city will become a chance for valorising architectures and settlement shapes, [then] a new semantic of the *open public spaces* [has] to provide the urban place with a new contemporary role. A city with cultural and functional boundaries, re-located by a soft and creative planning action, has higher probabilities of being recognized as a contemporary city which takes advantage of the polycentric creativity of enterprises, families and institutions. It is then the idea of setting against the present trend of building new infrastructures, which favours the birth of urban locks separated from the context, the chance of building new interconnected archipelagos, by avoiding overemphasising the centre or the centres as unique gravity poles.
>
> (Persico, 2013: 13–15)

Given that the urban territory in the communitarian dimension can be called *domesticable*, then "in the age of immateriality, *art* takes the shape of a place of highest cognitive and communicative, as well as, obviously, aesthetic processing" (Purini, 2003). This leads to the concept of design as very different from the conventional one. First, the street culture becomes part of the process, in which "the contemporary art and youth's culture is almost equal, and is as one of the paradigmatic incubators of use processes" (Ilardi, 2007).

A number of fundamental aspects of regeneration should be taken into account during the design procedure. For example, biodiversity, which is "the natural variety found in living systems and individuals at different levels" (Wiegand, 2003: 97), requires the achievement of various aims. These could be the improvement of links between tourism, business, and the environment, the provision for residents and visitors of a safe environment for participating in activities such as walking, jogging, and cycling, the guarantee for residents and visitors of opportunities to see the site aspects and to enjoy them, and the contribution to the region's economic regeneration (see Vidokle, 2003). The innovative regeneration requires various actions, for example:

> To think how it will be possible, and with which tools, to draw landscape...
> to identify possible routes to run so as to make – not only in some protected

paddock, but everywhere – this territory habitable and friendly...If today we can manage without engaging our *tekne*[1]...in order to find again and recognize possible landscape zones ...if it is the *tekne* which creates hazard to our habitat and our *logos*[2], then we should recognize that without the *tekne* we will have neither habitat nor *logos*...Even the recent sage recalls to *learn-from-nature*, the invitations to re-run the rules and the settlement principles processed by the European culture, can get effectuality if faced to natural reject (difference and identity) which is between us and nature, between landscape and nature. ... I like to see the [nature] that surrounds us as a shell and a mould which is modelling us and which is also modelled by us. [Finally the designer's role, which is that of making a project, should remember that] project [is] as a *refusal of oblivion and of the read*, [is an] ethical movement for saving the world and nature (nature as a memory of the world: what it is there and what it has been) against the evacuation of the reality and of the sensible presence of our body.

(Isola, 2003: 53–5)

The design process should be integrated with other expertise:

What we need today is a systemic approach which will integrate the bioclimatic researches on the relationship between the architecture morphology and the climate...with the biologic researches on the relationship between architecture and man...and his health, comfort, psychological and social aspects, while he lives in inner spaces, in order to consider the building as a living organism and the city as a *biotype*.

(Allen, 2008: 3)

Within this ecological vision, although we cannot expect that these interventions eliminated the agent that had produced the deepest effect on the urban shape, all the new interventions could at least be limited and new options could be promoted "for a different order, or a parallel one, which will allow the re-introduction of lost rhythm of the urban life" (Ingersoll, 2003: 43). This will only be possible:

[If] the architect [will] do shadow on himself so as to lighten the relationship between the individuals, discovering their immaterial relationships, their will of the other city, and helps to recognize the physical transformations needed for making needs and emotions come alive, by distancing the hypothesis of a forced community which usually ends up going down the wrong paths.

(Persico, 2013: 11)

An alternative vision of the design is supported by the concept of the *third generation city*, which has the following definition:

The organic ruin of the industrial city, [which is] true when the city recognizes its local knowledge and allows itself to be part of nature. [It is] like a weed creeping into an air conditioning machine; the industrial city will be ruined by rumours and by stories. The common subconscious will surface to the street level and architecture will start constructing for the stories – for the urban narrative. This will be soft, organic and as an open source based media, the copyrights will be violated. The author will no longer be an architect or an urban planner, but somehow a bigger mind of people. In this sense the architects will be like design shamans merely interpreting what the bigger nature of the shared mind is transmitting.

(Casagrande, 2008)

g The engineering role in regeneration (energy, information, water, and structure)

The new procedure for designing and completing any transformation currently needs to be supported by a great number of experts and tools. These include those contributing to the application of scientific knowledge, such as engineers, which could help with achieving a sustainable design for city regeneration.

g.1 Energy

The energy question, already faced as state-of-the-art requirements for the European 2020 pact, should be integrated into the design stage. The experts, while trying to reduce fossil fuel consumption, have discovered a number of technical solutions capable of allowing buildings and other constructions to employ less energy, exploit renewables, and guarantee more comfort with fewer costs.

During the urban regeneration process in particular, the provision of thermal, visual, and acoustic comfort requires a different approach. Nonetheless, it is possible to guarantee conditions in open spaces where citizens can feel protected from extreme climate conditions, such as strong winds, hot sunshine, and heavy rain. This will be possible with the synergy between bioclimatic design and innovative tools drawn by the plant-system technical engineers. For example, in order to guarantee external spaces with an acceptable thermal comfort during summer, the peculiar hazard of the heat-island effect can be taken under control with a careful design and a planning reprogram of the area for regeneration. This is done with the employment of those renewable systems that best adapt to the location and to the bioregion. Usually the energy experts define the homogeneous zones by means of detailed tools and through agreement between public and private institutions and then program a number of urban requalification interventions to guarantee appropriate conditions of livability and comfort.

Specific goals can be identified with the strategic solutions for the rational use of energy and for the energy saving, for the employment of the renewables in the residential, tertiary, industrial, transportation, public lighting and security, and the information sectors [of the city].

(Buono *et al.*, 2014)

The productive and settlement processes of the city can be referred to as the *ecosystemic quality*, as suggested by the Italian Minister for Public Jobs, which declares the frame as "the whole of conditions for guaranteeing upon time the inhabiting comfort in the cities...with respect to the existing ecosystems...by saving the use of available natural resources" [*Guida ai Programmi di Sperimentazione* (guide to the experimental programmes); processed by the Italian Law Decree of the Public Job Ministry, 1997]. Some urban solutions preview, for example:

[The] application of photovoltaic elements, with strong formal and technological innovation, with clear design and functionality components, and utilized as a partial or integral substitution of the traditional built components...and should satisfy the supply energy requirements also in the context of peculiar historic, artistic and landscaping values.

(Buono *et al.*, 2014)

Frequently, these projects and proposals present a great potential of diffusion and of reproduction and scalarity, which can be included in a *coordinated action plan*, with the following goals:

Acquisition of electric and combustion supply utilities, and of statistical data on energy consumption by productive and transportation activities...diffuse geothermal; popular actions; anaerobic digestion; public-private green-lighting; photovoltaic landscape; high efficiency mobility; energy exchange; energy efficiency provision for industrial machines and generators...experimentation of the wave exploitation...These could contribute to the achievement of the fundamental aims of the reduction of CO_2 emissions, of energy consumption, and the reconversion of the present energy asset towards solutions which involve a rational use of energy and a gross development and use of renewables.

(Buono *et al.*, 2014)

A number of easy solutions can be proposed in the public spaces, so as "to increase the thermal comfort or better, to reduce the discomfort in the hotter hours during summer days in the urban areas, [such as]...to enlarge the zones protected from solar direct radiation and covered by intensive green" (Selicato and Cardinale, 2012: 330–1).

g.2 *Information*

Any process for regenerating the cityscape relies on another kind of energy, which is represented by the following:

Expression, emotion, information and knowledge, and on their possibility to flow freely, and to leave evidence of their…history to be transformed into accessible forms of awareness, wisdom, insights, enlightenment and performance. In current times, much of these energies assume digital forms. We have learned to use mobile devices, social networks and other ubiquitous forms of communication to work, collaborate, make decisions, express our feelings, learn, communicate, establish relationships, and consume. It is, thus, possible to define…a *Third Infoscape*. Where the First Infoscape refers to information and knowledge generated within nature; the Second Infoscape refers to the information and knowledge generated in the industrial city (the second generation city, the city of infrastructures, of transactions, of sensors…); and the Third Infoscape refers to the information and knowledge generated through micro-history, through the progressive, emergent and polyphonic sedimentation onto the city of the expressions of the daily lives of city practitioners.

(De Certeau *et al.*, 2013)

Re-inhabitatory consciousness can multiply the opportunities for employment within the bioregion. New re-inhabitatory livelihoods based on exchanging information, cooperative planning, administering exchanges of labour and tools, intra- and inter-regional networking, and watershed media emphasizing bioregional rather than city-consumer information could replace a few centralized positions with many decentralized ones. The goals of restoring and maintaining [the local characters] invite the creation of many jobs to simply undo the bioregional damage that invader society has already done.

(Berg and Dasmann, 1977: 401)

In order to allow a more efficient fruition of the existing facilities under the urban viewpoint, the information experts state:

It could be desirable that the open data…as well as the private ones were available to all the possible interested actors (public and private institutions, companies and citizens). Therefore the modern information and telecommunication technologies[3]…for an integration-management [have]…the main goal of treating the land data, with regulated procedures and of enfolding them according to an open and standard frame in which they can be exported, interoperated and used by part of the community through various typologies (web, mobile, desktop) and by means of services designed according to the adopted standards…for example it could be possible to realize solutions such as the *smart-culture-and-tourism*, the *smart mobility*, the *last-mile logistic*, and so on.[4]

(Mazzeo and Fasolino, 2014)

These ideas, together with a central management of the information data, can represent a solution for city regeneration with no use of new construction and raw materials and products. This is because they will only create a network of existing built elements on the urban territory, whilst at the same time, improving and valorizing both the historic and the landscaping presence in town.

g.3 Water

The question regarding water comes from the need for people to get a supply of this resource; in fact the first settlements were born in humid zones. After humans had discovered many ingenious stratagems, cities settled in dry areas where water could even be supplied from far sources or extracted from rain or the soil.

> In the past, cities and towns have been established in areas that had secure water and energy supplies and fertile lands for food production. The burgeoning population growth and expansion of urban centres worldwide has placed increasing pressure on potable water supplies, energy and food supplies and the ecosystems services on which the community and the livability of the community depend.
>
> (Bristow, 2014)

Nowadays, during regeneration procedures, water can take a different role, becoming a parameter for designing in a more social way, in which the water itself, along with the energy, matter, and soil, can be better exploited to save precious resources.

The question of water in the city can be considered according to the activities run there, but also to the main sources of supply. The question itself can be divided into three main aspects: the provision for citizens, the dangers connected to the presence of water in the urban territory, and the attention to flowing waters, whether purely coming from precipitation or from urban uses.

The first issue, the pure water supply, has to be faced with the aim of achieving guaranteed water for everybody and of saving it from being wasted. This problem can be solved in a simple way: by recycling any kind of liquid in which water is present as much as possible and acting at urban level. The local administration, while managing the water supply for the citizens (where the water company is public), should find a way for centralizing the question and making strict regulations, mainly for large consumers, such as factories or hospitals.

The second problem, the hydrogeological hazard, is usually due to the following main events:

> Alluvial phenomena; collapse phenomena; rapid-flow phenomena; lift phenomena of under-soil hydric stratum and their pollution. Each of them has a specific danger score,[5] which shapes, on one hand with the vulnerability and on the other hand with the value, the hazard as a medium economic loss expected for a certain number of years.
>
> (Pianese and De Vita, 2014)

Usually, old European cities present a number of problems, such as those of the sewer network, which is often insufficient to gather and carry the exhausted waters, and the sinkhole of some territories due to the alluvial phenomena generated by the internal erosion in the superficial pyroclastic stores or due to losses or inefficiency in the hydraulic ducts.

> The rapid-flow phenomena of detritus or mud in the piedmont areas...represent a hazard source for the urban zones [here located]...mainly for the catastrophic effects due to their impact against structures and houses...this hazard is mainly due to the dense urbanization of land and for the continuous proliferation of the squatting building action.
>
> (Pianese and De Vita, 2014)

Some of the solutions identified by experts can be integrated into the urban design, and thus, become a potential for regeneration and valorization. For example, the volume of water reversed into the sewer—following heavy rains—can be reduced, and the eventual cavities produced by the previously mentioned phenomena can be employed for a number of activities after an appropriate consolidation and security provision. In this way, the water, as well as the soil, can be saved. As far as the question of the rapid-flow phenomena of detritus or mud in the piedmont areas is concerned, some solutions have been identified by experts in structural interventions (known as LLD or low level development) aimed at reducing the superficial flowing. Examples are green roofs for buildings, small tankards in courtyards, or basins in the parks (which could become also ornamental). Other solutions that tend to reduce the water content into the hydric receptors in valleys are included in the well-known techniques of the lamination basins (see Cimorelli *et al.*, 2012).

The third topic, the wasted water, has to be faced with the same intent as this book. Any kind of recycling system, or potential for different uses, should be pursued to create a network of different emissions in the soil that contributes to mitigating the pollution in the water tables. Another question, which is also important as far as water wastage is concerned, is that of waterproof ground. With current street cleaning, fluids cannot filter, and most of the time evaporate or converge into specific spots, while leaving the rest of the ground completely dry. Different kinds of asphalt already exist, which should be employed for any regeneration-designed system. Moreover, another important solution relates to minimizing the interruption of permeable surfaces and reducing the pollution of the wasted-water flows.

In regeneration processes, water engineers could help in designing solutions for saving the hydric resource, while architects and the energy experts could preview systems for enhancing potable water systems. For example, urban fountains (which can be closed, rather than being left running) or systems for checking dissipations due to being in a place that is uncomfortable (with too much heat in the summer). Moreover, some systems for recuperating rainwater in public spaces could be promoted by the synergy between planners, engineers, and local administrations.

g.4 Structure

Another fundamental role of scientific applications in the urban regeneration process is security. Not only can single buildings be put under security to avoid collapse but also all the city areas and systems can be as well. There are many hazards in the territory and a number of combining impacts. Examples are cascading effects, which can create situations of multiple risk; therefore, in order to guarantee good quality of life and security to people and their facilities, both in the residential and working areas, infrastructure, as well as the built heritage, should be studied and protected, so averting eventual economic and social reverberations due to any disaster. In particular, as far as the urban areas are concerned, experts have stated:

> It will be necessary to develop multiple-hazard analysis when the bioregion is affected by the contemporary occurrence of various events, which are independent, such as tectonic earthquakes, hydrogeological phenomena, coastal storms, volcanic eruptions, etc.…For example a seismic tremor could activate coastal storms, landslides, small collapses, liquefaction or constipation phenomena of ground; a landslide can generate overflow of natural or artificial rivers; a coastal storm can produce erosions, flooding or other environmental disasters; a volcanic eruption can cause earthquakes, lahar or tsunamis.
>
> (Landolfo *et al.*, 2014)

In the city, the inhabiting density of the urban territory and the historical built heritage can greatly affect many of these events. Therefore, security experts play a fundamental role during the regeneration processes of the city.

h The contribution of material science

One of the fundamentals of the bioregionalist design is clearly identified with the selection of suitable materials; therefore, a scientific approach to material studies is needed for calibrating regeneration technologies. New systems have been developed lately, and experts who can control the whole frame of performances are needed. Not only structural and technical energy engineers, but also chemical engineers, who have the role of understanding both the science of materials and the application of innovative technologies. Some of these new systems, already mentioned (see Part I, Chapter 3b), can be useful for city requalification because they have been developed either from recycled products, or from materials with a very high level of naturality, therefore, they are very sustainable. Only a few years ago, any artificial products had a very high level of polluting effects and a very low level of lifecycle-assessment values. The chemical engineers can also contribute to find intelligent solutions to transform urban wastes into usable products for designing the sustainable city.

Material scientists and engineers are specialists who are usually engaged in the research and design of materials. They usually define a number of criteria according to which the various materials are to be selected during any kind of project, which includes urban regeneration. An example is the service conditions, which indicate the properties of the material, in comparison with the requirement of the destined activities. Further criteria are the decay potential of the material and its consequent durability and economic evaluations.

> The more the designer…has familiarity with the various characteristics and with the relationships between the structure and the properties, as well as with the processing techniques of the materials, the more he will be able and sure in the sage choice of materials according to these criteria.
>
> (Callister, 2007: 5)

Experts are needed when dealing with so-called advanced materials. These are "usually shaped with both conventional materials, with enhanced qualities, and inedited materials with very high performances and recently developed…Some of these advanced materials are the semiconductors, the biomaterials…the smart ones and the nano-materials" (Callister, 2007: 11).

Another field of interest for material science and its experts is the assessment of the environmental impact of the material production and use. In order to take control of the various polluting effects, such as transportation and the energy requirement, the decrease of the weight of any object (either an engine or a building) is used as a solution or the systems for saving energy, for improving the efficiency and for employing renewables. Some of the materials are specifically destined for modern technologies to control air and water pollution, and "the existing toxic emissions of certain fabrication processes require to take into account the impact due to their disposal" (Callister, 2007: 13). According to material scientists, when a material comes from a nonrenewable resource, thus not able to be regenerated in brief times, for example, oil-derived polymers and some metals, a number of actions should be undertaken. Examples are "discovering [a] new supply, developing new materials with the same performances but less impacts, and/or increasing the recycling measures by processing new techniques" (Callister, 2007: 13).

If the material experts study both the production and the employment of the technologies, then their role in the urban regenerating process is also included in the knowledge of the various materials from inorganic ones, both ceramic and metallic, towards the polymers and composites. During the city transformation design, information on the materials to be used should be learned, such as the physical and chemical properties, the potential to be processed and synthesized, the structural performances, the appropriation to employment, and the eventual hazards which may impact on the natural and cultural environment: the last point is fundamental when dealing with a historical city.

Chemical engineering experts can contribute to environmental protection and pollution mitigation, while at the same time developing new systems for producing energy from renewables, for example, the algal biomass and waves, which represents a great potential in coastal cities.

However, the chemical and material experts, in synergy with the geologist, can also help during the eventual requalification of ancient materials, such as stones, bricks, the rammed earth, and timber, when strong decay occurs in the historical city.

i Restoration and requalification

Given that northern and central European cities, as well as Mediterranean towns, are very often based on an ancient settlement, and still preserve some of the old characters and historic values, the contribution of restoration and recovering expertise and criteria are to be taken into account during the regeneration process of the city itself. How the restoration specialists interface with bioregionalism is evident from the same definition of the bioregion, in which cultural identity is included (see Part I, Chapter 1b). However, it can be fundamental when the goal is that of recycling rather than constructing new elements in the town: the conservation is in fact one of the main criteria of any restoration act, aimed at safeguarding the natural and cultural landscape.

As the restoration "identifies the methodological moments for the recognition of the art work in its physical consistence and in the double aesthetic and historic polarity, so as to transform it in the future" (Brandi, 1977), then the regeneration of cities that are full of human history cannot avoid in taking into account such expertise. The Athens Chart clearly favors the following:

> Conservation of art and history monuments [declares that]…for safeguarding the masterpieces in which the civility has found its highest expression and which appear to be threatened…and for avoiding hazards…[it is necessary to adopt a]…regular and permanent maintenance, apt to ensure the conservation of buildings. In the case in which the restoration happens to be indispensable following the decay or collapse, it is recommended to respect the historic and artistic work of the past, without changing the style of any age. The conference recommends maintaining whenever it is possible, the occupation of monuments which can ensure the vital continuity, provided that the modern destination will be such as to respect the historic and artistic character…which will ensure in this manner a right of the community against the private interest.
>
> (The Athens Chart, 1931)

The strong fundamentals of restoration expertise have recently been included in the framework of a completely inclusive law in Italy, which clearly expresses the role and techniques to be adopted when dealing with cultural goods. The updated regulation (code 2004), which disciplines restoration works is considered as the main tool for such experts. It can clarify the role of these specialists within the regeneration processes for the city, wherever ancient, valuable, or landscaping parts are present. Starting then from *article 2* (Law Decree, 2002), in which the cultural heritage is defined as "made up by cultural goods and cultural landscapes, [then] mobile or immobile things which present artistic, historic, ethno-anthropological, archivist and bibliographic interest," the city parts are by force meant to become protected by this legislative tool and then subjected to some restrictions. However, the open spaces are mentioned as the "landscape goods, [that is] the immobile things and the areas as indicated expression of historic, cultural, natural, morphological and aesthetic value of the land and territory." The importance of the community in the processes of regeneration is also underlined by the same article that declares: "the goods of the cultural heritage belonging to the public are destined to fruition from community in accordance with the requirements of institutional use and always without being against the safeguard reasons."[6] The role of experts in restoration, within the urban transformation, is better defined:

> The valorisation of the cultural heritage consists of the exercise of functions ... and activities aimed at promoting the knowledge of the cultural heritage and at ensuring the best conditions for use and public fruition of the heritage itself, by different people, with the aim of promoting the development of culture. It also includes the promotion and sustainability of interventions of conservation of the cultural heritage.[7]
>
> (Law Decree, 2002)

As the ownership of these cultural goods is clearly public, the studies and projects here can be managed by the same entities ("state, regions or other public territorial institutions and also ... any other public or private institutions, including ecclesiastic;" article 10 (Law Decree, 2002)), which usually govern the city transformations.

According to the aims of this book (i.e., urban regeneration with a sustainable and bioregionalist approach, with waste, recycling, and a high level of naturality technologies), the restrictions due to the protection tools such as this code are important. This is mainly when there is discussion about forbidden interventions: they identify the actions that prevent the goods themselves being "destroyed, deteriorated, decayed or destined to those uses which are not compatible with the historic or artistic character, or such as to create prejudice to their conservation" (article 20 (Law Decree, 2002)).

Another important issue present in the restoration code, which creates the interface with sustainability goals, is the need for processing the environmental impact assessment.[8] Through the tools for evaluating the negative effects and the pollution, it is then possible to select, by the designer of the urban regeneration, the appropriate actions. In the code, the latest are considered as means of conservation and are then defined and finally classified as "studying, prevention, maintenance and restoration."[9]

However, what actually places restoration and urban regeneration on the same plan is the goal of valorization. The code defines

> "the activities at valorisation of the cultural goods [as] the stable constitution and organization of resources, structure or networks, the availability of technical or financial expertise and tools, aimed at the exercise of functions and at the pursuit of this law's goals." Other common aims are public involvement, participation, and social subsidiarity.[10]
>
> (Law Decree, 2002)

Restoration experts have also included the landscape in the cultural goods to be protected. Therefore, the concept of cityscape can be studied under the viewpoint of restoration actions. In fact, the code "safeguards the landscape as far as those aspects and characters are concerned, which identify material and visible representations of the national identity, as expression of cultural values."[11] It indicates the tools for restoring (or we could say regenerating) the landscape itself, such as the landscape plan that includes a number of actions very similar to those identified here as part of the regenerating process; in fact, article 143 (Law Decree, 2002), referred to as the landscape plan, identifies the following processes:

1. The recognition of the territory object of planning, by means of analysis of its landscaping characteristics, printed by nature, by history and their interrelations; 2. the recognition of the immobile things and areas declared of high public interest, with their delimitation and representations at the right scale, and the definition of specific use descriptions; 3. the recognition of areas with peculiar characters, with the appropriate scale for their identification and determination, and the identification of use-prescriptions aimed at ensuring their conservation; 4. the definition of a number of measures, necessary for the correct insertion and integration in the landscape context of interventions for land transformation with the aim of pursuing a sustainable development of the interested area.

(Law Decree, 2002)

Also for landscape protection, the items of participation are included in the restoration expertise, together with the diffusion, to increase the knowledge of the cultural built heritage.[12]

j Economy and finance

The role of economics and the social sciences in the design procedure for regeneration processes can demonstrate that bioregionalism can create benefits to citizens and welfare to workers if the synergy between the various stakeholders is active, continuous, and transparent. One of the different economic theories that can be recognized in the updated and innovative movements is *de-growth* (see Chapter 2). It has been lately outlined and already has a number of followers in various fields of society members, not only in the economic sector but also in the construction sector, in the social sciences, and in others. The idea, coming out from all the people involved in processing this model,[13] can be summarized in the few words that Latouche, the very well-known French economist, had written in one of his various essays:

The de-growth is not a negative growth. It should be better to talk about *a-growth* as it is talked about *a-theism*. In fact, it deals exactly with the abandon of a faith, a religion, that of finance, progress and development. It is by now recognized that the never-ending searching of growth is not compatible with a finite planet, while the consequences (producing more and consuming less) are on the other hand very far from being accepted. However, if there will not be a route inversion, an ecological and human catastrophe is expected for us. We are still in time for imagining, peacefully, a system founded on a different logic: that of a de-growth society.

(Latouche, 2012)

Other forms of alternative economic frameworks have been proposed around the world to create a civilization-management system closer to human needs and desires.

Some territories had met a proper decline and others, built around cultural diversity, habits' endurance and quality of artistic, scientific and cultural research, had succeeded. What has affected this revolution? The capacity of interesting the young talents and of valorising their ideas, focusing on values, is the element that bonds all these experiences of success. The resource of major value in order to be competitive is the *creativity*…where the creative stay, richness, occupation and life quality are recorded…It is possible to create a blend between artistic creativity and enterprise creativity…[because] the immaterial component in the goods results always fundamental; the advanced economic systems should focus less on the use-value of products and more on the symbolic and evocative valence which the goods and the experiences of services express and tell.

(Casoni *et al.*, 2008: 164–5)

The role of sociopolitical and economic expertise within the architecture can be proven by the fact that very often these issues become part of the preliminary stage of the project.

Studies on topics taken from sociology and psychology, economy and ecology, mathematics and communication theory…the only subject which, paradoxically, is missed is architecture…[as]…the concrete means an architect uses to solve the tasks he is facing, that is architecture forms…we need more architecture [as an answer to these anthropic issues], not less.

(Norberg-Schulz, 1968: 257–8)

In substance, [a different way of seeing the city questions had brought to reflect on the fact that] economic goods and common goods, both specific and un-specific, should be born or be shown. In the past the communities had demonstrated harmony with the land and environment, and to be able to save the resilience of the territory … today, the maintenance of the future is the absent concept in the [city] project, almost as if the environmental conditions, the Nature's design, and the various levels of naturality, had not subjectivity and incisiveness on the time and the space of inhabiting.

(Persico, 2013: 12)

k The sustainability experts

With the hope that "the term environment from the green ghetto [could] be integrated into cultural discourse" (Jacob, 2009: 25), architecture, planning, engineering, and all the other actors, as well as their disciplines, should collaborate and cumulate their efforts to create a sustainable team of experts. They should be able to think of ecology as part of the natural process of building and transforming both in the small and the larger scale, such as the city. In order to enhance such expertise's role, the question of *space* in architecture should recover its value, rather than destroying its significance. In fact what had happened "from Paul Reuter's carrier pigeon, to the automatic ticket and the first transatlantic telegraph cable, is that the increased speed of communication has brought with it a *dissolution of space*" (Jewson, 2013). There are no longer distances in our small world, and nowadays "this has physical implications on a territorial and architectural scale" (Jewson, 2013).

If we start to think again about city architecture as a means for orientating life towards more sustainable welfare, and stop thinking of the world as a machine, with no limits and no identities, then we can give a start to a more human earth. For achieving this goal, it is necessary to create "a synthesized understanding of post-industrial environments in a fully matured manner, a kind of *ecology-as-conceptual-art*, where what we normally think of as architecture is just one part" (Jacob, 2009: 25). The sustainable experts will become those who tend to carry the identity of places, natural cycles, human needs and desires, communities' requirements, the biosphere, and atmosphere and hydrosphere conservation, within the fundamentals of the city design and transformation procedures. These specialists can then help any time it is necessary: the architects, engineers, planners, historians, restorers, public administrators, scholars, and any other expert involved in the city regeneration, should include the environment and bioregionalism in their goals and ensure, by means of the assessment tools (see Part I, Chapter 3d), that these goals are achieved.

l The technology of architecture during the urban regeneration process

Technology, "the invention of the artificial machines which mark the beginning of the technique,"[14] is the study and procedure with which any transformation is applied to a number of fields, for example, the territory, buildings, and the city. It is then obvious that technological factors affect the regeneration processes in the measure in which the transformations are to be applied with material and concrete actions and interventions. What kind of technologies can be considered useful for a sustainable city is decided only following a deep knowledge of their potentials, performance, and impact-provision (see Part I, Chapter 3b and c). The experts of technology have the role of superintending the process to take control of the selection of technologies, materials, and products to be employed: from here is the importance of the technological approach to the environmental design.

What is the contribution of technological choices during the entire lifecycle of the city transformation? By making a comparison between conventional large-scale design procedures and those grounded on bioregionalism and sustainability, it is possible to identify an answer. The first stage of any design procedure with the technological approach is the analysis. The technology experts require any possible information data, with as much scientific reliability as possible. The technical knowledge of the site and its performances and characters have to become the main elements that push the decisional process towards an appropriate solution. The kind of scientific investigation that is needed depends on the main goals of the regeneration that the city requires.

The consequent actions and interventions to be run are in accordance with these investigations. By maintaining, for analogy with the studies run in the field of architecture technology, the distinction in natural and cultural, these regeneration actions for studying the territory should begin from the natural presence and processes, which are taken back to the three categories of essential data. Soil and undersoil morphology, vegetation, and climate conditions, which correspond to the same amount of dynamic processes, are switched on by the combined action of various environmental elements, as are ground landslides, hydro-geological failure, and coastal erosion, which are included in the category of the *edaphic* factors. The growing processes and development of seasonal vegetation belong to the second order of data, green spaces.

Solar radiation, humidification phenomena, thermal performances and daily and seasonal swings, lighting and acoustic conditions, variations in air quality consistency, and water pollution and contamination, are parts of the third kind of data, hydroscopic and environmental data.

The potential of possible design transformations are greater than conventional modern systems happen to achieve and exploit the following categories of data. For the first category of data, ground morphology, it is possible, and often necessary, to apply artificial processes such as earth movements, soil consolidation, and other modifications. For the second category, that of green spaces, there is very often the requirement of reforesting, and various techniques of naturalistic engineering are aimed at shaping new processes that could connect the natural action of the environment to the transformation required by the territory. Finally, it is necessary at any scale, from the territorial to the building, to the single object, to interface the climate elements with a design procedure that is aimed at harmony with the type of weather. This is by means of passive collection of the sun and wind, switching on water recycling processes, and the use of light and humidity, through a check of phenomena of oxygenation and decontamination of air and acoustic processes.

In conclusion the role of technology experts during the regeneration process is that of checking the availability of the resources, reliability of the scientific investigations, and the evaluation and deep control on the materials, products, and any kind of concrete and physical solution that should be applied to city regeneration. In particular, during the regeneration of the city, technologies (see Part I, Chapter 3c) should be (as said) soft, with low ecological footprint, re-entering as much as possible within the natural cycles and employing few pure resources (matter, water, energy, and soil).

However, the main role of the technology of architecture for city regeneration is that of integrating any of the choices (energy saving devices, low-tech materials, water use, and so on) into the design procedure. This proceeds with an *ex-ante* and *ongoing* process of evaluation of any possible solution to achieve the previously mentioned aims of sustainability, bioregionalism, and de-growth.

Notes

1 *Tekne*: art in ancient Greek.
2 *Logos*: discourse in ancient Greek.
3 These technologies for the standard management of data are cloud systems, the software architectures aimed at facilities, the standard interface for the access to data, and facilities.
4 The main levels for these solutions are the following: interface service level, data access enablers, data integration, data domain sources, the transversal level which implements the functionality of the network for the monitoring, the management, the supervision, and the integrated security (Mazzeo and Fasolino, 2014).
5 The danger score is defined as the temporal probability of occurrence of a certain calamity phenomenon.
6 The safeguard is defined in article 3: "the safeguard consists in the exercise of functions…and is based on an appropriate knowledge activity aimed at defining the goods of the cultural heritage and to guarantee the protection and conservation for public fruition."
7 Article 6 continues: "as far as the landscape is concerned, the valorization includes also the re-qualification of the buildings and of the areas under safeguard, which are compromised or decayed, or the realization of new current and integrated landscape values."
8 Article 26 declares in fact that "for the project of works to be undertaken by environmental impact assessment, the authorization is given only if there is an environmental compatibility."
9 Article 29 (conservation):

> prevention is [defined] as a complex of activities able to mitigate the hazard situation connecting to the cultural good in its context; the maintenance is the complex of activities and interventions aimed at the control of the conditions of the cultural good and to the maintenance of the integrity, the functional efficiency and the identity of the good and its parts; for restoration it is intended the direct intervention to the good through a complex of operations aimed at the material integrity and to the recovery of the good itself, to the protection and the transmission of its cultural values.

10 Article 111 continues:

> private subjects can participate and co-operate to these activities. Valorization can be public or private. The public valorization is shaped according to the principles of free participation, of subjects plurality, of continuity of exercise, of treatment

> equity, of the economy and transparency of the management. The private valorization is a socially useful activity and can be recognized as a social solidarity action.

11 Article 131 states:

> landscape is intended as a territorial expression of art identity whose character is derived from the action of natural, human factors and from their interrelations. The safeguard of landscape is aimed at recognizing, safeguarding and, when necessary, recovering the cultural values which it expresses. The valorization of landscape concurs to promote the development of culture.

12 Article 144 states:

> within the procedures for approving the landscape plans a number of things should be ensured, such as the institutional consultation, the participation of the interested parties and associations which bring diffusing interests, defined by the in force regulations for the environment and environmental decay, and by white shapes of advertising.

13 As far as the de-growth model is concerned, a great number of scientists and economists have lately been dealing with that. See Glossary.
14 "The technique is that [of] proper…art and craft; the whole of the rules which superintend the art or the craft" (Mercier, 1984).

Part II

Technological approach to environmental design

Application to some urban case studies

Introduction

The second section of this book tries to show how some of the theoretical and methodological criteria, announced in Part I, could be practically applied to urban areas.

Among the various proposed strategies and design examples, some solutions may be useful for future cases in European cities with similar troubles and conditions. The open spaces in the city represent the target of a large part of the actions required for regeneration, as they create the feedback for any kind of requalification, being the natural context of city life.

> An open space is a complex braid of social, cultural and economic activities within urban districts, answering to their peculiar logics, just because it is inscribed in that context. It is the simultaneous expression of social realm and physical system, with the microclimate and the physical components of the space. The exchange between the various elements and actors takes place only in the sites of open life and nowhere else.
>
> (Carbone, 2009: 72)

How the project can improve these spaces depends on a number of factors, including the designer's sensitivity. It is possible that the designers "deliver, to the community, open spaces of undoubted appeal…often weak of attractive strength, since small care is due to the interactions which are established between the great number of components" (Carbone, 2009: 72). Open spaces are either in central areas or on the outskirts and they should be studied as much as possible until all is known about these components. In Clément's *Third Landscape* (see Part I, Chapter 5e), he states the following:

> The necessity of training our gaze into recognizing and understanding the Third Landscape. This requires a new possibility for vision and knowledge dissemination in urban natural environments, a renewed sense of aesthetics, and a morphological sensibility for the potential of interaction and communication offered by our surroundings.
>
> (Clément, 2004)

The boundaries of an urban area can even be understood according to this concept, and so defining the end of it as soon as the third landscape starts.

All the solutions presented here are generally aimed at reducing the amount of constructed elements in order to limit the ecological footprint and safeguard the identity of urban architectural places. The solutions are also very often oriented towards employing materials and products that present a high level of naturality, that are recycled, and that come from waste. The design applies mostly flexible, transitory, and very soft technologies to upgrade their present performance and regenerate the urban open spaces without actually *generating* a completely different city.

6   Urban squares

a The requalification of Mercato Square in Naples

Following a change of the primary function (a general market) of the Mercato Square in Naples, and a subsequent abandonment by the users of the external areas, as well as of the existing and important historic buildings, the square has been subjected to long-term neglect. It presents some peculiarities, in its shape,[1] in the closeness to the sea, and the presence of an important church.

The proposed solution[2] focuses on the following topics. *City regeneration* with a rediscovery of the original commercial activities, an attentive respect with valorization of the historic memory of the place and its identity. The *Technological approach to environmental design* by means of the employment of the natural, recyclable and zero km materials (such as the *Pallet*[3]) and the use of photovoltaic panels. Various *experts* are needed for the completion of the project, such as planners, transportation engineers, and architects. Among the useful *scientific investigations*, the analysis of solar radiation and frequency, wind development during the various seasonal changes, studies on the planning situation, and a number of other investigations were done. They aimed at being conscious of the existing contextual situation before selecting the design strategies and knowing which parts are to be regenerated and with which modalities. The project previews the insertion, within the square, of a number of so-called *all-around kiosks*, which during the day can be used for the market activities, while in the evening, by sliding with their wheels along the linear guides, can get closer to each other to create a gallery (Figure 6.1).

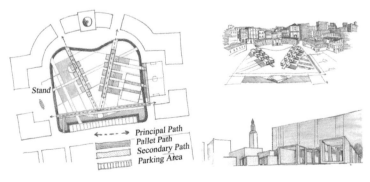

Figure 6.1 The project for Piazza Mercato: the plan with the main paths and views

b The municipality square in Sant'Antonio Abate (Naples)

The square, Vittorio Emanuele II, in Sant'Antonio Abate, which is surrounded by municipality offices, a modern church, a few buildings with some shops, and a bus stop, has been the object of an urban program by the Town Hall administration. It aims at enhancing the space and providing some places for possible events and/or community activities. The approach of the project has been that of the bioregionalist regeneration, by means of preparing the square to host future activities rather than by means of edifying new constructed elements.

The local authority, the university, and the mayor promoted a small competition with the previously mentioned brief among the young student architects, who had responded with a number of solutions. At the end of the competition, an exhibition with the participants' solutions was held in the square itself, where the citizens could hold discussions to promote the *participation process* during the decision procedure for the completion of the regeneration.

One of them previewed the requalification of the area facing the municipality building with an ideal recovery of the original path that linked the city of Sant' Antonio Abate to the close town of Lettere.[4] Moreover, in a space that is drawn in the plans as a leaf shape, which captures the very ecological sense of the proposal, the areas dedicated to events are located either in the open air or covered by small and transitory roofs with photovoltaic panels.

This project has been mindful of the *city regeneration* topic with the recovery of the existing original path and the insertion of a stage for citizens' events. It also considers the achievement of the *technological approach to the environmental design*, with zero km materials such as timber, limestone, and volcanic stones. Some renewable energy sources were exploited by means of the *solar tree* system. A variety of *expertise* was needed for the proposal, such as planners, transportation engineers, and architects. The *scientific investigations* consisted of the climatic elements and the environmental phenomena, useful for gaining evidence of the existing situation and in helping to select the design strategies (Figure 6.2).

In another solution proposed by the students, a number of places for relaxing and resting with the use of benches and sitting systems alternate with the use of trees, which can guarantee shade in those areas with great solar irradiation. Two sheltered zones were proposed, covered by timber and vegetation, a new space alongside the church and another substituting the existing space at the bus stop, which will create a better milieu for those waiting for a bus.[5]

The following topics of the sustainable approach to urban bioregionalism are *city regeneration*, by means of the requalification of the square that will allow the daily as well as the nightly use of the place. *Technological approach to environmental design*, with the selection of such lighting systems that will serve as provision for visual comfort, but also as elements that can shade some zones during sunny days and as a main focus of the project; in addition, wide employment of the autochthone vegetation. The synergy between *expertise* such as planners, technical engineers, architects, technologists, and sustainable consultants was helpful for the selection of the materials, techniques, and the whole frame of the

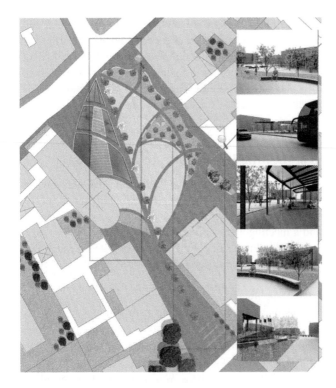

Figure 6.2 The first project for the Sant'Antonio Abate municipality square: a general plan and renders of the sitting areas, the new bus stop, and green areas

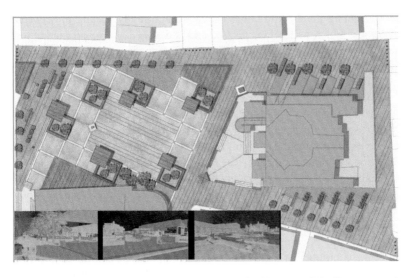

Figure 6.3 The second project for the Sant'Antonio Abate municipality square: plan and renders of the sitting systems

conceived design. Some *scientific investigations* included both the cultural and the natural environmental elements, such as the apparent solar path during the year, the main wind directions and frequency, and the acoustic and air quality for the existing situation before the design proposal, in order to answer adequately to the actual needs and the local conditions. In this case, the environmental analysis has enabled the understanding of the place's disadvantages, such as the large and uncomfortable nonshaded zones (Figure 6.3).

c The requalification of the Milli Egeminlink area in Istanbul

Immigration represents one of the main problems of this international city. In order to create a suitable *regeneration* of the park, besides being exploited for the usual conventional activities (children playing, picnics, meetings, etc.), a congregating spot is created, and a meeting area for the multicultural market. The market is one of the main symbols and areas of activity of this city and through this employment, a better integration between different populations could possibly be triggered.

The project[6] previews the requalification of the park with the inclusion of some aggregation zones, linked together by means of various paths. Routes with twists and turns to avoid right angles, which do not favor interaction among people: the organic figures coming out from this can become a symbol of the city of Istanbul, which presents itself as a very chaotic place.

This path will be completed in wood and intersected by sitting areas, as well as eating and drinking zones in between different bended lanes. Some pergolas can help the users to protect themselves from the very hot sun in summer and at the same time from the wet days in winter. A certain amount of vegetation connects the area of meeting points with another area, still currently vacant, in between other buildings, which will be recuperated and is destined as a small market for immigrants. The connection between the two areas is made up with some bended lanes that are repeated with the same curves as within the market zone.

The use of natural materials such as wood for the paths, glued laminated timber for the pergolas, and of products at zero km—such as the porphyry cobbles—demonstrates a *technological approach to environmental design*, as well as the use of natural fabric for the tensile structures employed for the roofs of the exposed areas. The main intervention of the whole design is the requalification of the floor surface. The latter needs to be improved because it is impermeable and abandoned, like much of the area. The project proposes a more permeable system also aimed at avoiding hazard of subterranean troubles; however, it will be treated to eliminate reflection, which often occurs in Istanbul on sunny days. The choices could contribute to cooling in the summer, so improving people's comfort and reducing the wind inconvenience in winter. Therefore, a number of *scientific investigations* have been made in order to create the consciousness of the context: solar and wind analysis, together with the planning situation have helped the selection of trees and bushes for protection from the wind. Planners, architects, and transportation engineers,

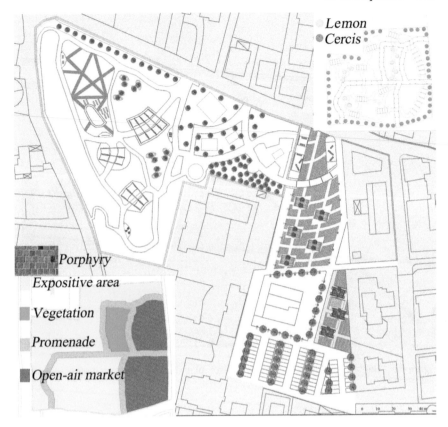

Lemon
Cercis

Porphyry

Expositive area

Vegetation

Promenade

Open-air market

Figure 6.4 The square of Milli Egeminlink in Istanbul: design plan, concept lines, and vegetation selection

in addition to the landscape *specialists*, could help to achieve the goals of the project during its future completion (Figure 6.4).

d The small plazas in the historical center of Naples

The requalification of small parts of the city is an emerging issue of contemporary regeneration, especially when a historical center is present. The proposal described here[7] previews a number of temporary actions for Santa Maria la Nova plaza in Naples, including the employment of very light and multifunctional sitting areas, which could be dismantled, and the application of the *street art* technique aiming to enhance the artistic and picturesque character of the place. The main principles of the applied bioregionalism were *city regeneration*, for achieving both commercial activities and architectural heritage in which they are enhanced to discourage vandalism, the squatting scooter park, and the neglected green areas. The *technological approach to the environmental design* included the use of natural materials at zero km such as *maritime pine* for the timber uplifted footrest. The involved *expertise* for

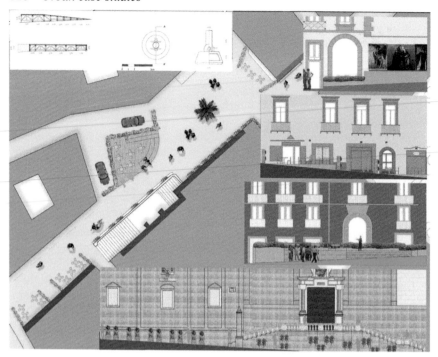

Figure 6.5 "Santa Maria la Nova" square project: plan, facades, and detail of the pergolas

such a project should be the architects, the transportation engineers, the planners, and the historians to program the rearrangement of the area and the surroundings. This case needed deep *scientific investigations* to understand the cultural background of the existing milieu and the urban regulations, as it is located in very ancient and historically stratified zones, but also to analyze the natural environmental factors, such as the sun, wind, temperature, and the noise milieu (Figure 6.5).

Notes

1 The shape resembles a symmetric pentagon.
2 This is part of a degree thesis in the Department of Architecture, University of Naples "Federico II," by the students, G. Angieri and R. Assorto.
3 The Pallet is a waste-derived element from the fruit storing system, all in timber.
4 This solution was processed by the students, C. Esposito and G. Gargiulo, Department of Architecture, University of Naples, "Federico II."
5 This solution was processed by the students, M. Giacca, S. Ceglia, and G. D'Argenio, Department of Architecture, University of Naples, "Federico II."
6 The project has been conceived within a convention between the University of Naples "Federico II," and the Istanbul Technical University, with the main work of the architect E. Adamo, tutored by Profs. Yidliz and Francese.
7 This solution was processed by the students, M. Chianese, E. Ferrazza, C. Grieco, V. Iorio, Department of Architecture, University of Naples, "Federico II."

7 The oblique city

Introduction

The question of the oblique urban districts is strongly represented in the cities where the natural morphology of the ground presents hilly areas and various slopes. The historic solutions here were usually established by an intricate texture in which the staircases, tilted streets, and the little undulating squares became part of the context. Any project carried out in these areas should consider, within the framework of regenerative actions, the enhancement of the morphology of the ground, and the role of pedestrian areas. Other important elements, often present in these oblique parts of the city, include landscape, which adds value through the panorama and the derived privacy and the chance of employing green areas, which is high due to a low number of vehicles.

a The Pedamentina in Naples

The case presented here is a peculiar path made up with a long staircase and a variety of morphologic developments and urban textures. It has already become the object of interest by various institutions, which led to listing the area as UNESCO protected heritage. The importance of the place, and the fact that it is very near to an important Neapolitan castle (Sant'Elmo) and to an ancient religious building (Certosa di San Martino), its popularity with tourists, its neglected state, and the need for better security, besides a general decay, have created the demand for regenerating and requalifying this urban system.[1]

Within the university program, together with the Municipality of Naples, represented by the mayor, who came to take part in the exhibition of the output projects, with the Pedamentina association, which represents the inhabitants' desires and point of view, a number of solutions have been presented. They were aimed at reducing the existing problems and enhancing the potential of the zone, while promoting eventual *participation* design procedure.

One of the solutions regards the recovery of some portion of the oblique district by starting from an important issue, that of information necessary for knowing and then enhancing the Pedamentina for both tourists and citizens. One small temporary information point could be located at the top of the staircase, where the castle is, and another, with the shape of a sheltered arcade, at the other entrance at the bottom. The refurbishing of the stairs and some zones for resting are previewed, with a more comfortable path, aimed at valorizing the fantastic views of both the near agricultural landscape, still present in the area, and the farther landscape, which embraces both the sea and the magnificent Vesuvius volcano skyline.[2]

The main principles targeted by the project for *city regeneration* were: enhancing the panoramic spots and requalifying the important historical part of the city that connects the superior residential districts to the central historic texture. *Technological approach to environmental design* is shown in the use of an innovative system, which is in the shape of a tulip and is mainly made up of a polylactic acid material,[3] and could gather rain water (Figure 7.1). A number of *experts* are needed, such as botanic and landscape specialists, technical engineers for water collection and plants, architects, and restoration experts. *Scientific investigations* are needed for comprehending the cultural and the environmental situation, such as the historic texture of the buildings surrounding the staircase, the sun, wind, and other climatic factors, and the presence of noise at the top and at the bottom of the Pedamentina, where this fantastic and silent zone meets the chaotic city streets (Figure 7.2).

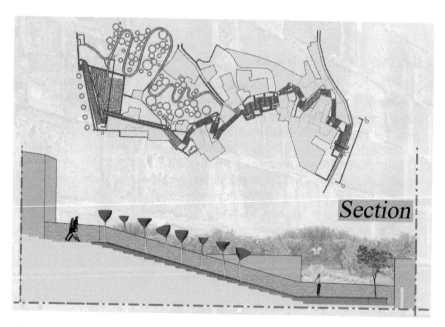

Figure 7.1 The project for the Pedamentina: plan and section with the red tulips system for water collection

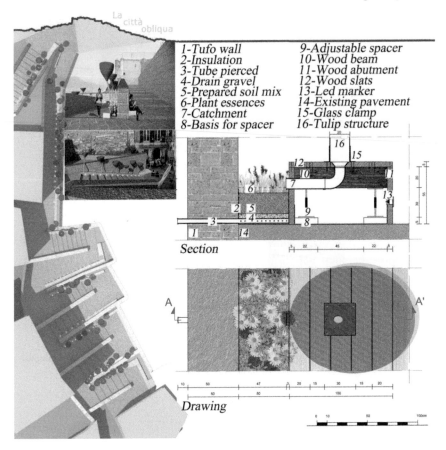

1-Tufo wall
2-Insulation
3-Tube pierced
4-Drain gravel
5-Prepared soil mix
6-Plant essences
7-Catchment
8-Basis for spacer
9-Adjustable spacer
10-Wood beam
11-Wood abutment
12-Wood slats
13-Led marker
14-Existing pavement
15-Glass clamp
16-Tulip structure

Section

Drawing

Figure 7.2 The project for the Pedamentina: general plan, views, and details of water drainage

b The Pendino di Santa Barbara (Naples)

One of the small-sloped paths in the historical center of Naples, the Pendino di Santa Barbara, has been the object of an interesting proposal, which is part of a research carried out in collaboration with Spain, Romania, and Portugal. The place, connecting the lower *decumanus* of the ancient urban settlement with the central upper one, is mainly made of stairs, and completely closed in between two curtains of tall stone buildings. After a number of *scientific investigations,* such as temperature, shading, acoustic considerations, wind, and cultural landscape and popular traditions, a number of critical aspects emerged. For example, the fact that the whole area throughout the year is mostly in shade, the difficult access for some people that creates problems of security, and the absence of a specific destination (Figure 7.3).

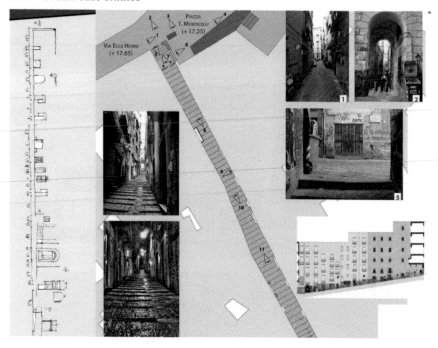

Figure 7.3 Pendino di Santa Barbara. Requalification of the staircase: plan, façade, and views

Therefore, the proposed project has tried to reduce these problems by enhancing the potential of the area, that is, the important facades, the presence of a traditional small kiosk for selling water, and the humidity levels, which are good for planting some green spaces. The latter was one of the main objectives of the proposal, which could create a sort of arcade, which makes the place more livable and fresher in summer. Another element of the project was the improvement of the ground floor entrances, by promoting new small commercial activities and designing new doors. The *technological approach to environmental design* has pushed the designers[4] to select many ecological materials such as timber for the doors, the green supporting system in bamboo, and the rehabilitation of the ancient water kiosk in fir wood: the latter can be dismantled to show the ancient marble portion of the original bar (Figures 7.4 and 7.5). The required *expertise* for this project is made of the architect, industrial designer, planner, restoration specialist, and the botanist. The difficult path of the oblique part of the city can be *regenerated* if the application of this project could be completed with the consideration of possible solar collection through a number of devices located on the top of the green arcade.

Figure 7.4 Pendino di Santa Barbara: concept and kiosk design

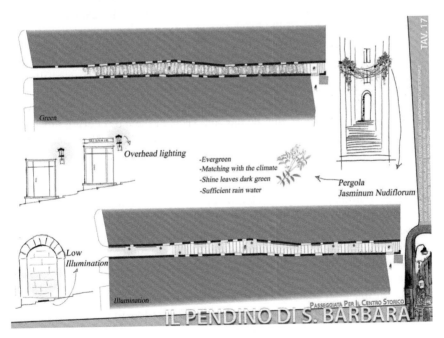

Figure 7.5 Pendino di Santa Barbara: the green arcade design

c The stairs and paths in Positano (Salerno)

Another case of the oblique city is that of a very famous little town in the well-known tourist area of the Amalfi coast, Positano, which in ancient times was built on a very hilly slope towards the sea. This slope morphology today causes significant problems in the town, as the small unique road is not adequate for the large flow of tourists during spring, summer, and the autumn. The proposed solution, in accordance with the Town Hall Authority, which had highlighted the need, is the creation of a small square, at the top entrance of the long hilly path, which actually coincides with the city itself. This square could host tourists as well as the local inhabitants during their wait for the bus, for friends, or for organized trips. It is a very busy area, but does not offer the necessary facilities and guarantees of adequate viability. The proposed project, besides requalifying the under-located building, is meant to create this new space, with a number of facilities and plain zones for resting and organizing other activities.

The main functional areas, proposed by the preliminary project's solution, are a green area with trees, an information point, a small scene for events, a staircase, capable of being dismantled, with a covered roof for the public attending these events, a small fountain, a flower bed, an equipped bus stop, and a fence for protecting the visitors viewing the splendid panorama. The latter is previewed as a sort of filter in the shape of a small wall that recalls the original historic parapet running along the entire long provincial street of the Amalfi coast. It is provided with squared and multishaped holes that permit the visualization of single spots of the panorama at a time, with a number of superior arches in timber as support for the visitors. In the bus stop area, it becomes higher and covers those waiting for the bus. It is here, in the small plaza that the dismantling platform with the staircase for the audience is located. It also satisfies the requirement of privacy, by hiding the square from the provincial road, and it is covered with a flexible texture tensile structure for protecting the spectators from the sun during the day and keeping them dry during the night. Through the small plaza, it is possible to proceed from the main street to the very sloped stairs that lead towards the inferior height of the street itself, so connecting the two levels via a pedestrian path.[5]

Some principles of bioregionalism are applied. For *city regeneration*, the small plaza could promote a new convivial area for running a number of activities. The selection of green areas, the timber for the fence and tent structure, and the rammed earth for the small information point construction match the *technological approach to environmental design*. The required *expertise* for the project will be architects, engineers, city planners, and the municipality technical team. A number of *scientific investigations* were necessary, such as the position of the solar shading systems and protection from the wind, which is prevalent from the northwest. These were useful for locating the shading elements, such as the tensile roof and the trees to facilitate the various activities (Figure 7.6).

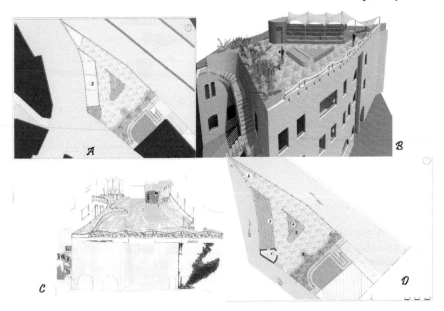

Figure 7.6 Positano square, regeneration project. (A) The plan at roof level (1, information point with garden roof; 2, translucent tensile structure roof); (B) tridimensional view; (C) preliminary design sketch; (D) ground floor plan of the plaza (1, information point in rammed earth; 2, green area; 3, fountain; 4, timber scene; 5, staircase with timber structure and finishing in rammed earth; 6, bus stop; 7, stone sittings)

Notes

1 The presented ideas were also part of an exhibition held under the supervision of the Mayor of Naples in March 2014.
2 Designed by C. Ciccarelli, C. Piscopo, V. Quiriti, and R. Rossi, tutored by Profs. G. de Martino, F. Mangone, and D. Francese, 2014.
3 Polylactic acid (PLA) is a material obtained by the fermentation of natural materials with arid presence (such as hemp, corn, or sunflowers) via multiple condensation.
4 This solution was conceived by G. Ballirano, V. Borrelli, and A. Coppola, tutored by Profs. G. de Martino, F. Mangone, and D. Francese, 2014.
5 The project is part of an ongoing process of cooperation between the Town Hall and a private owner. It is designed by the architect D. Francese and the engineer R. Rinauro.

8 Urban waterfront

a Caracciolo Avenue in Naples

The serious question of the Neapolitan main waterfront of Via Caracciolo would require an entire essay, but in this case, the important thing to underline is the fact that for a long time this street has been acting as a barrier between the seafront and the city, for the presence of cars and parking areas. Even though this seaside is located very close to an important urban public park (Villa Comunale), the link between the green area and the sea was, and in part, still is, mediated by a very congested, vehicle accessible, road. Only a small pavement allows the citizens and the tourists to enjoy the wonderful view of Vesuvius and the Neapolitan Gulf. Recently, the municipal administration had opened some of this route to pedestrians, therefore, deviating the vehicular traffic elsewhere. Following this functional variation, the whole waterfront needs a careful redesign and rearrangement to become a real urban space.

One of the proposed projects takes the wavy motion of the sea as a basic concept for the requalification of this waterfront, so giving birth to the bicycle path and pedestrian route, and to a number of resting activities such as restaurants and the meeting areas (Figure 8.1). The geometries of the small square located in the middle of the street (Rotonda Diaz) were also changed according to the idea generated by the sea.[1] The main criteria of the environmental approach for *city regeneration* in this project are the need to generate new pedestrian qualities and to enhance the relationship between the sea and land. For the *technological approach to environmental design*, there is the employment of recycled timber for the paths and sitting areas and the introduction of new green areas, which are integrated with benches. The required *expertise* is made up with planners, transportation engineers, architects, technical and hydraulic engineers, and historians. *Scientific investigations* were used for the comprehension of questions related to natural and cultural processes, such as the sun path, dominant wind directions, air and water quality, and the local cultural landscape.

Another proposal for the same question of the central Naples waterfront is the introduction into the area of a new cycle lane with the appropriate additions. It will be a substitute for the existing path, which is provisory and not yet adapted to all the requirements, especially during dark hours and is thought as a continuous path through the whole city. Another zone has been designed with sport activities with open-air facilities, such as tennis courts, volleyball pitches, children's games, and a path with a length of 100 meters for jogging along the seaside.

Figure 8.1 Via Caracciolo waterfront, first proposal: plan, view and section of the regeneration of the square "Rotonda Diaz"

At the western side of the two paths (for jogging and bicycles), an information point and a bicycle hire office are proposed, where a wide parking zone allows the citizens to leave their cars before entering the pedestrian area. The path for walking is located in between deciduous trees for shade in the summer, with different colors indicating the various activities, and with interruptions to provide space and visual interest to those who are walking. The facility area is defined by a path through a number of arcades that can be equipped for various events, such as exhibitions, theatre, and a local district market.[2]

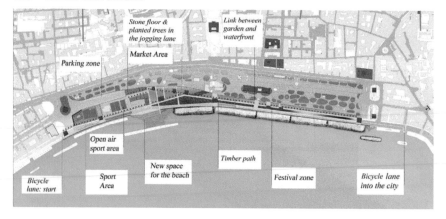

Figure 8.2 Via Caracciolo waterfront: plan of the second project's concept

As far as the *city regeneration* is concerned, the entire axis along the sea is meant for providing services and facilities, now absent, and providing the chance of experiencing the area in a different way. It will not only be a walking and secure zone, but a free area towards the sea with no barriers, for sitting and a leisure space. Moreover, the access to the sea is improved and the number of passages is increased. The *technological approach to the environmental design* is clearly demonstrated by the employment of recycled materials, such as for the street and pavement zones mostly finished with an artificial stone made up with waste from the site works; but also by the use of photovoltaic panels. Many *experts* are needed such as transportation and hydraulic engineers, useful for rearranging the traffic routes and the sea front, planners, for helping to meet the local administration's standard tools, and architects. A number of *scientific investigations* have been made for providing the guidelines for the proposed project such as the natural and cultural environmental analysis in order to be conscious of the present situation and proposing a conceptual design appropriate to the site itself (Figure 8.2).

b The San Vincenzo pier in Naples

Another important area of the Neapolitan waterfront is the historical, "Molo di San Vincenzo", a very long pier, entering far into the sea, which has for many centuries been a symbol of the harbor area (Figure 8.3). The proposed solution for the requalification of this very decayed but still valuable infrastructure previews the employment of a number of bioclimatic strategies, which allowed the selection of various techniques and drawings according to solar and natural ventilation performances, analyzed during the *scientific investigation* stage. The number of existing arches, which represent the actual structural system in the local *tufa* stone, have been exploited as part of the project, instead of building new parts while still housing new activities that could revitalize the area. The main *city regeneration* goal was that of enhancing the place as a witness of the historic background of the pier itself and of the whole area near the port. The *technological approach to the*

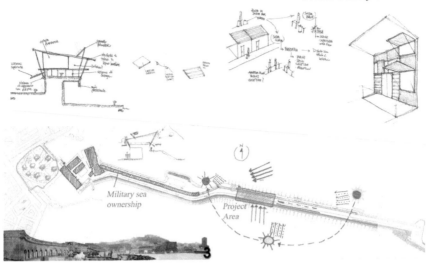

Figure 8.3 San Vincenzo pier: project sketches, general plan with solar access, and view

environmental design was achieved by selecting ecosustainable materials such as wood, recycled elements, and the flexible structural elements that are easy to erect and dismantle and do not require water. For this important reason, the presence of some structural and hydraulic engineers is required, with some interior designers for the complex division of the very small internal spaces, obtained by the areas under the arches, and other *experts* such as planners and historians to help during the integration with the ancient city parts.[3]

c The requalification and adaptation of Haliç (Istanbul)

The area of Haliç, close to the sea, but opposite the more touristic waterfront at the northeast, where the historic central zone of Istanbul can be found, is part of the same wide municipal territory of the city, which has recently been neglected, often even hosting some waste landfill. Although it presents some opportunities due to the number of green areas, the closeness to the sea, and the presence of a nearby traffic-congested area, no planning programs are currently applied to the zone.

The proposed project previews the demolition of the existing fence that runs along all of the perimeter of the area (20,900 m^2) and a new use destination, a natural park, which is one of the main requirements of the city. The geometrical background of the whole drawing of the park is due to a number of axes coming from the main direction of the existing area on the west side (the sea is on the east), which become the more important timber paths. Intersecting with those axes are a number of secondary lanes proposed in stone, which represent the only trace in between the green, which dominates the whole area (Figure 8.4).

The *city regeneration* goals are clear from the obvious improvement of the historical center of the city and the chance of enjoying an area that had been long secluded and fenced in. The *technological approach to the environmental design*

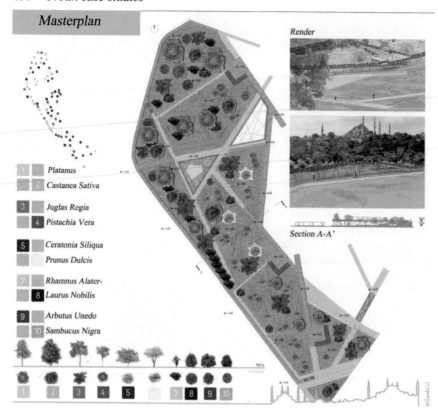

Figure 8.4 The regeneration of the Haliç area in Istanbul: general plan, section, vegetation distribution, and views

has applied the choice of ecosustainable technologies, such as natural materials and the green parapet for the above ground paths. The required *expertise* is made up of the architect, landscape specialist, botanical professional, and planner. The *scientific investigations*, such as the solar and wind paths, planning and other studies, aimed at being conscious of the existing contextual situation before conceiving the design strategies and selecting the hazard zones, have been made with a number of site visits, direct surveys, and photographic shoots, as the base of the project.[4]

Notes

1 The work is part of the degree thesis in architecture by C. Guarnieri, Naples, tutored by D. Francese and aided by E. Adamo.
2 Designed by P. Velotto (2013), Department of Architecture, University of Naples, "Federico II."
3 D. Agretti and R. Avino: degree thesis in the Department of Architecture, 2013, tutored by D. Francese and P. Giardiello.
4 Degree thesis in architecture by F. Orefice, 2015, tutored by D. Francese, F. Uz, and G. Longobardi.

9   Decayed suburbs

Introduction

The question of suburban areas is substantial; however, some solutions can still be proposed, starting with the idea that

> even though the [existence] of a worldly economic power in the metropolis…
> there are places in which the poverty is concentrated. In the physical shape
> of the city this is expressed in the decayed districts in which the immigrants
> from other countries live in non-admissible life conditions.
>
> (Moccia, 2012: 185)

> The UN have underlined the question of slums and proposed a solution. In
> the strategy for the implementation of the Millennium Development Goals,
> number 7, target 11, they propose "by 2020, to improve substantially the lives
> of at least 100 million of slum dwellers, while providing adequate alternatives
> to new slum formation".
>
> (Moccia, 2012: 187–8)

In order to achieve this goal, three action lines have been carried out: physical, social, and political recovery of the slums; development of the city economy and livability; and country development with a balance between nature and polycentrism.

a The Scampia park (Naples)

This Neapolitan district had sadly become known all over Europe for a number of problems, such as social and environmental decay. The original settlement of residential buildings had not been followed by the provision of any facilities, infrastructure, or links to other parts of the city, thus creating very difficult living conditions that had led the inhabitants towards criminality and drug dependence.

Therefore, the proposed project[1] was meant to guarantee at least one of the nonsatisfied requirements of the place: the green park. An area destined for such a function, although identified for such green functions by the local planning tools, has currently neither been designed nor built. The requalification of this

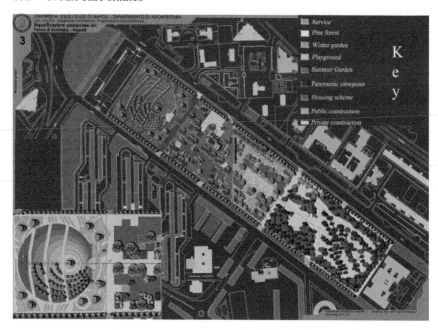

Figure 9.1 The regeneration of the "Scampia" park: general plan and detailed winter garden plan (bottom left)

semi-abandoned area is proposed with the insertion of a small hill with a panoramic viewpoint, a summer garden, a winter garden, some water games, some pine trees, and a number of useful facilities (Figure 9.1). The *city regeneration* goals could be achieved during the requalification actions when the abandoned and neglected sites can be given back to the residents, so stimulating the movement of people from other parts of the city. In addition, enhancing the value of the whole district of Scampia. The *technological approach to the environmental design* is clear, for example, in the selection of the certified ecological timber or in the recovery of the embankment aimed at creating the panoramic viewpoint hill. The *scientific investigations* were useful for achieving the green selection, according to sun exposure, the eventual wind protection or enhancing, and for any other choice as an answer to the environmental performance of the place, but also for defining the various activities for future users. Many *experts* will be needed for the completion of this project, such as planners, landscape specialists and botanical gardeners, architects, and geotechnical engineers.

b The outskirts of Marrakech (Morocco)

The proposed design experimentation is peculiar both for the selection of the use destination (an institution for developing the education for youths on the rammed earth techniques) and for the employment of the same matter (rammed earth) to a number of constructed parts.

The project is aimed at the requalification of an outskirt district of the city of Marrakech, in Morocco, which was completely decayed and abandoned, but located in a very crucial zone of the city itself. The area can be regenerated by limiting the construction to few buildings, while significantly rearranging the large surfaces, all of them defined by a peculiar shape, coming from Islamic art: the hexagon.[2] The main goal of the whole idea is that of valorizing this marginal area as a means of *regenerating* the entire ancient Moroccan city, while displaying new common spaces. Moreover, by employing natural materials and local technologies such as the *pisé*,[3] the project aims at a *technological approach to environmental design*. For its eventual application, a number of *experts* will be needed, the botanic and garden artists for the rearrangement of the autochthone plants and green areas, transportation engineers for the reorganization of the traffic around the case study area, technical engineers for the comfort provision and the passive solar cooling systems, and material engineers for the control of the performance of the rammed earth. In fact, the modern way for treatment of the rammed earth previews a less artisanal and more checking system to ensure the optimal levels of health and comfort. The architects are needed for the completion of the main idea and for the management of the complex spaces and their levels, mainly the small library and open courtyards. A number of *scientific investigations* are needed for the application of the proposed preliminary project, such as climatic analysis, for the control of very hot summer radiation and to identify solutions for protecting the inhabitants from the wind, which is dominant at times during the year, mainly coming from the Atlas Mountains. In addition, scientific investigations are needed, such as the study of the planning tools so as to include the right destination into the proposal, and other analyses aimed at being conscious of the existing contextual situation before selecting the design strategies, and for knowing the parts to be regenerated. The whole investigation frame has led to propose open shapes that can allow good natural ventilation and more air dispersion to cool the spaces, both outside and indoors (Figure 9.2).

c The regeneration of the suburban area in Casoria

In the small town of Casoria, an industrial site was once located, not far from the center, which has recently been abandoned and neglected, so providing the surrounding suburban area with a very difficult and inconvenient situation. In the wider area ($98,344$ m^2), some existing buildings can be reused, some others can be valorized, while the external spaces have been completely left to wild vegetation and to anthropic decay.

After having analyzed the climatic, social, environmental, and physical situation of the whole area, with some *scientific investigations*, it has been noted that the potential for reuse was great. This is due to the good annual insolation (for the absence of close large obstacles and for the width of the place), the almost absent wind flows, except for some days during the summer, the need for improving the commercial services, the green spaces and sport facilities, and a general desire of the population for taking possession of the place. Therefore, the proposed project[4] for the *regeneration* of this suburban area was mainly meant as a general

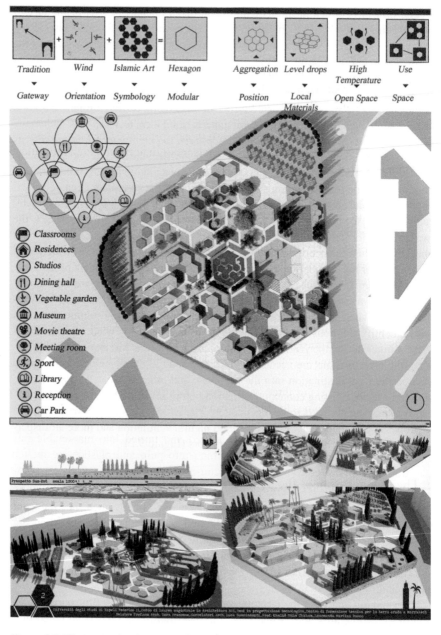

Tradition	Wind	Islamic Art	Hexagon		Aggregation	Level drops	High Temperature	Use
▼	▼	▼	▼		▼	▼		▼
Gateway	Orientation	Symbology	Modular		Position	Local Materials	Open Space	Space

Classrooms

Residences

Studios

Dining hall

Vegetable garden

Museum

Movie theatre

Meeting room

Sport

Library

Reception

Car Park

Figure 9.2 The center for the rammed earth: project with the concept (top), general plan, section, and views (bottom)

requalification, starting with the rearrangement of the whole ground surface with new floors in natural materials (local stone, earth, etc.). Then rehabilitating some of the existing fabrics, while providing some identity elements that could give some meaning of the place back to the inhabitants of the town, according to the bioregionalist approach (see Part I, Chapter 2).

The fence that surrounds the whole area will be transformed to guarantee access from the main town roads and the easy interpretation of the place's new activities and facilities, while conserving some memory of the existing old systems. The activities have been selected, according to the existing identified users' needs, as a big plaza in the central zone for the development of the existing industrial archaeological sites and for the distribution of the various other activities that will be located all around the plaza itself.

On the west side, a memorial park has been considered as a tribute to the workers and the activities once located in the area, which is partly made with vegetation and with some transitory timber walls. In addition, there is a general rearrangement of the floor with some written words, which the pedestrians (the only allowed category of users) can read. All around the plaza there will be a large portico that will follow the morphology of the ground to help the users reach the upper levels, while accompanying them in a shaded and protected pathway.

In the northwestern area, a theatre and some green facilities will be located. In the northern area, some sporting fields and facilities will be linked to a substantial park for children and adults. In the east side, an existing building will be reused for housing a covered open-air market, while on the northern access to the plaza, a new building will instead house a small commercial center. This building, besides following the layout of the portico and ending up in a green pergola, is the sole constructed element of the whole

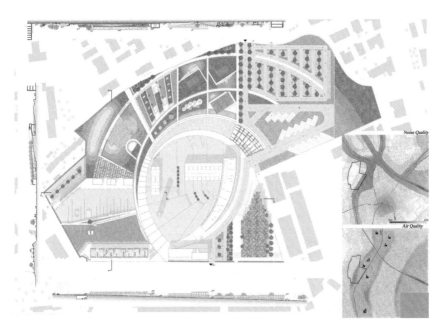

Figure 9.3 The regeneration of the suburban area in Casoria: general plan and sections, noise and air quality maps (bottom right)

project. It is proposed as a bioclimatic fabric, with naturally lit shops and corridors, a glazed surface on the south facade for collecting the sun's rays, as well as for converting them, partly, into active photovoltaic renewable energy (Figure 9.3). All these choices were aimed at a sustainable *technological approach to environmental design*. The roof of this building, which is structurally sustained by a very ecological timber structure (in a particular kind of glued laminated timber, employing a natural fix), has the shape of a shell and is partly opened to permit its use as a ventilated public place for the summer. Some *experts* are needed for the completion of this project, such as botanists, landscape specialists, architects, planners for connecting the surrounding urban area to the site, and some sociologists for interpreting the desires of the population.

d The market square in Sant'Antonio Abate (Naples)

A very decayed and neglected area in the town of Sant'Antonio Abate has been taken as a case study for requalifying a suburban zone, where a market and a small green park are located (Figure 9.4). The rearrangement of the street and of

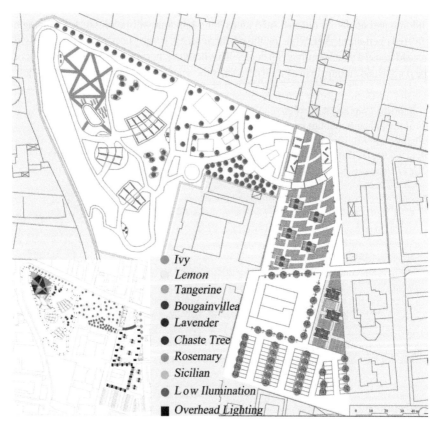

Figure 9.4 Sant'Antonio Abate market square: regeneration plan and greenery selection (bottom left)

the market kiosks is derived from the offset of the arch generated by the continuum between the path within the park and the main street. In the project, some resting areas are proposed for the park, built in timber, which can be useful both for visiting and for future temporary exhibitions, in particular aimed at an immigrants' market or events.[5] The goal of *regenerating the city* can be achieved with the insertion of very few elements in timber and fabric, which can be potentially dismantled and flexible, for the park as well as for the market area. Some natural materials, some zero km products, and the photovoltaic collection are selected for the market during the day and for other activities, such as eating, drinking, and other events during the night; this satisfies the *technological approach to environmental design*. Various *experts* are needed, such as landscape architects, material engineers, transportation experts, and planners. Many *scientific investigations* have been conducted and they focus on areas such as sun and wind analysis and soil quality, which helps with the selection of greenery, and useful for protection from the wind in winter and from the sun during summer.

Notes

1 The project is part of the degree thesis in Architecture by F. Canettieri, 2013, Department of Architecture, University of Naples.
2 Degree thesis in Architecture, Naples, 2013, by M. Russo, tutored by L. Buoninconti, D. Francese, and G. Longobardi.
3 The *pisé* is a peculiar traditional technique, which creates a number of layers of rammed earth thrown in stages to compose walls. This material, the stratified raw clay for shaping large walls, besides being a traditional technique by which the same walls of the ancient city are built, is also a very sustainable system. It is 100 percent recyclable and it is very healthy, good for comfort and energy-saving buildings, and does not require a fossil-fuel supply in the production. Finally, it does not emit any kind of polluting substance.
4 R. Sodano's thesis in architecture, 2015, tutored by D. Francese, A. M. Puleo, and G. Longobardi.
5 The project was undertaken by M. Adamo, with the help of G. Longobardi and D. Francese.

Part III

Technological approach to environmental design

Application to peri-urban case studies

Introduction

A brief introduction to the question of peri-urban areas is useful while focusing attention on the fundamental role of bioregionalist and sustainable technologies during the process of regeneration. A strong and truthful picture of peri-urban zones can be read in the following lines by Saramago:

> The landscape is dusky, dirty, does not deserve to be watched twice. Someone has given these wide expanses, with an aspect of everything rather than rural, the technical name of Agricultural Belt, and also, for poetic analogy, that of Green Belt, but the unique landscape that the eyes can see on both sides of the road, which covers, without perceivable solution of continuity, thousands of hectares, is made up of big fabrics…Under them, out of the crossing people's view, the vegetation grows.
>
> (Saramago, 2000: 4)

Peri-urban is a mixed term derived from ancient Greek and ancient Latin in which *peri* (Greek) means very close but also surrounding; and *urban* (Latin) stands for the product of a city. A more scientific definition of territory is one that describes peri-urban as follows:

> A space around the city stated by the agricultural areas in the proximity, in which the external boundary of the urban suburbs, the infrastructures, the great centralized nucleus of the commercial and productive places, the sprinkled texture of the living dimension, full and empty spaces, all concur to draw, in the varieties of combinations, a new territorial figure, typical of the contemporary age, generated by the cum-participation of spatiality and of a different compactness and morphological as well as inhabiting density, not necessarily identified with urban or rural society.
>
> Definition from the dictionary *"Desmots de paysages et Desjardins"*
> in (Ilardi, 2007: 39)

Identification of this area can clarify both the concrete layout of the places and their perception by users and citizens. "The peri-urban presents *un-waited spatial orders…its spatiality is not wanted by the actors who had contributed to produce it"* (Salvemini, 2006: 13)

It is clear that peri-urban areas derive from a "new spatial revolution unleashed by the birth of a society built on hyper-consumption...an open field in which energies, powers and new relationships are released" (Ilardi, 2007: 39). A spatiality produced by a distraction and by an excess of spaces, uses and society. Peri-urban is "a hazardous situation, a confused and contradictory condition," it is made up of "territories and societies that are not easily identifiable." Around the city, there can be zones that are part of a city, of a rural space or ecology, while "the peri-urban spaces draw the division lines between the urban and the extra-urban item" (Mininni, 2012: 13).

From a European perspective, peri-urban areas are often understood as mixed areas under urban influence but with a rural morphology (see Caruso, 2001). Other pertinent terms can be recognized in *inhabiting, suburbs,* and *marginality,* and the binomial term of private-public. "Neglected for a long time, the notion of peri-urban area has been introduced to describe the heterogeneous settlement pattern at the urban-rural interface, replacing the former model of an urban-rural dichotomy" (Errington, 1994). Peri-urban space is identified by Donadieu as an *urban country*:

> Peri-urban sites should make some effort to introduce – in the home, in the health, in the job – the new idea of comfort and welfare...It once again joins together and closes the last layers of the city and those of the country; it re-articulates thicknesses and densities of the remaining objects from the modern items, also those coming from the hardest and not solved suburbs: the so called urban fringes, where the country people are living a profound social change.
>
> (Donadieu and Boissien, 2001: 19)

Peri-urban territories are not completely neglected by people or by the local administration, but conserve in themselves some meaning, in the bioregionalist sense, according to which a number of various sensations can be stimulated in the people perceiving, living in or only passing by, these zones.

> They say that the [peri-urban] landscape is a mood, that the external landscape can be caught with the internal eyes...these extraordinary internal organs of the view could have not seen these factories and these hangars, these smokes which devour the sky, these toxic dusts, these eternal muds, these soot crusts, the yesterday rubbish swept away by the today rubbish.
>
> (Saramago, 2000: 82)

Conversely:

> In the peri-urbanity...there is the chance of perceiving...the nostalgia of a nature which is already all revealed, re-dimensioned and domesticated, but that can still astonish and help to be detached.
>
> (Mininni, 2012: 17)

Then the peri-urban subject can be considered rather as "a concept, perceived intuitionally rather than rationally, [in which] the spatiality of the peri-urban rises when a *reflective practice* generates knowledge" (Schon, 1983: 7).With this interest for the region, a variety of proposals could be promoted and a new project attitude can be introduced, just caring for the past existing presences, built or semi-natural, without applying new strategies and behaving as if there was a void. These peri-urban areas sometimes have potential and they are important for closeness to the city, as well as for the function of a buffer zone that they actually play. It is an area that has analogies with the ecotone, which, in ecology, represents an environment in tension. The definition is as follows:

> Every habitat located between bordering systems with an high diversity differential, the sea, the coast, the meadow, the forest, that, becoming close, would produce a "third-ality," i.e., the sea front, the marsh, the tree-full meadow around the forests, and so on.
>
> (Mininni, 2012: 18)

An ecotone has some *emerging properties*, created by nature in conditions of proximity. The ecotone is an environment in which the Darwinian fight selects the more resilient species.

The idea of the *Third Landscape*, introduced by Gilles Clément, can acknowledge these areas with new values and roles.

> The Third Landscape – an undetermined fragment of the Plantar Garden – designates the sum of the space left over by man to landscape evolution to nature alone. Included in this category some places are left behind such as urban or rural sites, transitional spaces, neglected lands, swamps, moors, peat bogs, but also roadsides, shores, railroad embankments, etc. To these unattended areas others can be added as spaces set aside, reserves in themselves: inaccessible places, mountain summits, non-cultivatable areas, deserts; institutional reserves; national parks, regional parks, nature's sanctuaries. Compared to the territories submitted to the control and exploitation by man, the Third Landscape forms a privileged area of receptivity to biological diversity. Cities, farms and forestry holdings, sites devoted to industry, tourism, human activity, areas of control and decision permit diversity and, at times, totally exclude it. The variety of species in a field, cultivated land, or managed forest is low in comparison to that of a neighbouring *unattended* space.
>
> (Clément, 2004)

Therefore, new projects and soft interventions are to be proposed for this "space around the cities, invaded by the urbanization, still made up by agriculture, [but] invested by a process of great innovation" (Mininni, 2012: 14).

How can an urban form of settlement, thought of as the least environmentally viable and the most socially segregated, be worthy of a project? Precisely by ceasing to think about it as if it were irrelevant in the urban field, disqualified by its own existence, unworthy of public policy except for the one that would stop it.

(Vanier, 2011: 2)

The innovative role of the peri-urban regions, in terms of comparison between the land use and the peri-urban spaces, is that the "economy is dematerialized" (Mininni, 2012: 17) and that the three sides of the question, city, peri-urbanity, and sustainability cannot be separated.

For the past forty years, society has become more and more suburbanized.... Actions taken by researchers in the so-called hard sciences (physicists, climatologists, ecologists), which warn of global change (energy, climate, biodiversity), strongly resembles the *last combat*: the so-called post-carbon; livable city will either be compact or it won't be at all. Urban sprawl – the most unacceptable version of suburbanization – should stop immediately.

(Vanier, 2011: 1)

Peri-urbanisation, as a process of the physical expansion of settlement areas but also socio-economic transformation, has been recognised as a major spatial development beyond the urban fringes.

(Zasada *et al.*, 2011: 59)

Here, a new line of design process is proposed for the peri-urban areas. What can be taken as potential for these territories is not only the social and cultural issues that still persist in the location (some old fabrics of a traditional or bucolic origin), but also the vegetable and organic raw material that can be taken as resources for the interventions. In addition, the production and application of a number of very soft, sustainable, and bioregionalist technologies (straw, cane, other plants, vegetable earth for adobe bricks, etc.) as well as the existing sun and wind natural energy supply are to be considered. The use of a number of byproducts, from rural life and still present in some peri-urban zones, can be incentivized and promoted, and they can become partially witness to the existing meaning of the site itself.

10 Tourist waterfronts

a The lake front of Bacoli

Not far from the main city of Naples and in the famous area of Campi Flegrei, so-called for the fertility of the soil and known for the still present volcanic action, the small village of Bacoli has lately become an important tourist location for both foreign and regional people. A great number of seaside areas are present and some archaeological ruins can be found in the ground and under the sea: from here, there is a need for valorizing the site and improving accessibility and usability. In particular, the great number of shore systems adapted to sea users and the various restaurants and public activities run all year but mainly during summer, while some areas are still abandoned and neglected by local authorities and tourists.

One of these sites has been the object of the proposal,[1] which intends to create a complex redesign, of which the waterfront represents the focus of the actions. The first issue to be performed by the design procedure has been that of creating and arranging some windbreak barriers, mainly made up of trees and bushes, and aimed at safeguarding and protecting the vegetation and agricultural zones that were previewed in the internal areas of the site. There will be an area for typical local produce growing, besides a smaller zone with some didactic vegetable gardens, aimed at reteaching the population to appreciate the benefits of natural growing (Figure 10.1).

The proposed area is to be arranged with sports facilities, pedestrian paths, and bicycle lanes. These and other solutions were intended to form part of a general *city regeneration*, in which structural requalification is the main action to be applied, according to the needs of functionality and usability of the whole area. The employment of zero km materials, which can be recycled, in addition to very soft and simple techniques, shows the *technological approach to environmental design*. The required *expertise* for this proposal is made up of planners for the legal and urban control of the solutions, transportation engineers for linking the superior area of the case site with the central part of the small town, botanists for the selection of appropriate vegetation, and finally, architects for the general management of the whole design procedure. A number of *scientific investigations* of the prevailing winds and wave direction during the year, solar radiation, and

Figure 10.1 Regeneration of the Bacoli lakefront: general plan (top left), detailed plan with green (top right), views (center left), and sections (down)

other environmental and technological analyses were fundamental for the choices aimed at locating the various activities for regeneration.

b The waterfront in the village of La Corricella

The well-known little village of La Corricella, located on a small island in the northwest end of the gulf of Naples, has been declared a UNESCO heritage site for its peculiarity. The Mediterranean style of urban texture, as well as the colors and intricate structure of the various fabrics, intersected with each other, has actually made true the original name: *Corri-cella*, from the ancient Greek (*corri* coming from *chorus*, i.e., village, and *cell*, deriving from *Kalos*, which means beautiful), which responds to the significance of "beautiful village" (Figure 10.2).

The village descends towards the sea, and ends up near the sea at a shoreline and small bay, where the fishermen traditionally landed their equipment, including fishing boats and nets. Now the sea is separated from the shoreline by a long rock barrier to mitigate wave action. The area can be enhanced by promoting an improved tourist destination.

The proposal is aimed at valorizing the space in between the seaside and the cliff by changing the role of the barrier into that of a filter (Figure 10.2). The project presented here (designed by the architects G. Paone, E. Pisapia and

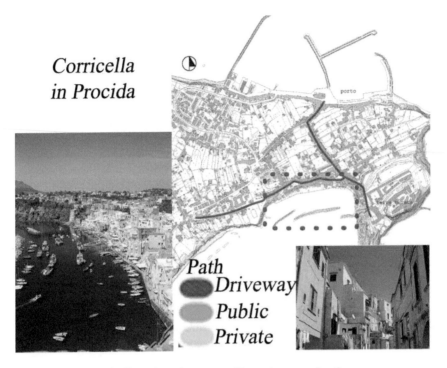

Corricella
in Procida

Path
Driveway
Public
Private

Figure 10.2 La Corricella project: the present village pictures and paths

Figure 10.3 La Corricella project: facade from the sea (top left), ideal connection between coast and sea (top right), and render of the "water plaza"

M. Valentino) previews the requalification of the waterfront of the village, mainly by reusing the ground floor space, presently occupied by fishermen's boats. The activities that could be improved and located in these rooms are a diving center and some small facilities for fishing and tourism. These spaces could be more easily accessed with the transformation of the existing staircase that links the upper part of the village with the shore by intervening on the systems and favoring vertical greening. However, the main core of the proposal is the drawing of a *water plaza*, which could reconnect the village to the cliff in front of it. This idea can create a real *regeneration*, in which the pier is no longer seen and conceived as a barrier between the sea and the earth, but rather as a system of multiple filters, with various activities. The selection of materials, all sustainable and bioregionalist, such as timber for the structural parts, natural fabric for the tents, and stones for the reef, witness the *technological approach to environmental design* (Figure 10.3). Some *experts* are needed for the completion of this project such as the littoral protection engineers, planners, and architects. A number of *scientific investigations* were aimed at being conscious of the very delicate situation of the area, the small size of the interventions, and the UNESCO site safeguard. In fact, the design concept was mainly due to climatic analysis, fundamental for the definition of the optimal exposition and the employment of natural resources such as the sun and wind. The sun for collection during winter and protection during summer and wind for promoting air movement under the tents.

Note

1 The project, made under the supervision of Prof. Francese and Arch. Massimo Scatola, was then partly presented for the discussion of the degree thesis in architecture of Luigi Nicola Petito.

11 Disused industrial areas

Introduction

An important aspect of peri-urban areas is the conversion of once-active industrial functions. Where some factories are in disuse, the areas are often neglected, sometimes decayed, and in some cases, greatly polluted. The following is known:

> [The] development of peri-urban areas involves the conversion of rural lands into residential use, with closer subdivision, fragmentation and a changing mix of urban and rural activities and functions. Changes within these areas can have significant impacts upon agricultural uses and productivity, environmental amenity and natural habitat, upon the quality of the hydric supply and the water/energy consumption. These changes affect the peri-urban areas themselves and the associated urban and rural milieu.
>
> (Bristow, 2014)

a The sustainable requalification of the industrial area in Emmerbrucke (Switzerland)

The project presented here considers that the environmental quality concept includes the comfort of people, sustainable use of resources, and the control of waste applied to architecture, which requires great care during all stages of the building process.

The proposed idea,[1] besides the erection of some new social housing, previews the *regeneration* of the site where there was once a factory with its various buildings and equipment (Figure 11.1). The transformation of the larger area is mainly designed to rearrange places with the *bioclimatic approach*, in which the selection of vegetation and the orientation of the resting areas can maximize the collection of sun during the long Swiss winters. In addition, it will try to create protection from overheating during the few summer days.

The *technological approach to environmental design* is also demonstrated by the fact that all of the materials employed in the reorganization of the open spaces are mainly local, natural, and in some cases, recycled from wastes derived from previous industrial activities. The selection of materials for pavements, streets, and the little squares where people meet, are always considered to avoid any soil impermeable action, as the original concept of recovering this area was exactly that of saving the soil. The selection of the trees follows the bioclimatic approach:

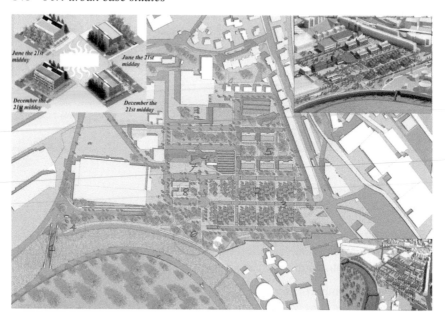

Figure 11.1 Emmerbrucke's regeneration: solar access to the houses (top left), views (top right and down right), and general plan (1, new bridge for linking houses and city; 2, pedestrian paths and urban courtyards; 3, streets at reduced speed; 4, bicycle lanes; 5, bio-pond; 6, green area; 7, the memory of the ancient factory; 8, service center with a kiosk, shops, toilets, parking, bus stop, and covered pedestrian area)

the areas that need to be protected from the sun in summer but to collect heat and light during winter are chosen with deciduous varieties. When it is required to provide protection from the winter and cold winds from the Alps (Phoehns), the trees are instead very dense and evergreen. Other choices were all aimed at saving energy, re-employing resources, water and waste, and safeguarding health, comfort, and livability of the whole area, as result of *scientific analysis* of the climatic and environmental situation.

Many *experts* will be required for the requalification of Emmerbrucke, such as transportation engineers for the rearrangement of the internal, as well as external, street access. Botanists and landscape specialists will be required for the protection and enhancement of the wider area. Architects and planners will be needed for the redistribution of the new activities. Also required will be bioclimatic experts for the selection of building orientation and a technological specialist for waste recycling and natural material employment.

b The sustainable regeneration of the disused area in Casalnuovo (Naples)

The project[2] previews the introduction of a green area just outside the town of Casoria, which is near Naples. A number of *scientific investigations* were aimed at selecting the specific variety and location of the bioclimatic devices: therefore, the sun, wind, sound, and air quality have been analyzed to create a useful database for design decisions, mainly aimed at solar and wind collection. In addition to the design of the autochthone vegetation species in the public open spaces, there will be a sporting area within the park, once used for a factory, and some pedestrian paths (all built with permeable materials and layers) that intersect with the cycle lanes. On the northern side of the whole area, some small new constructions are envisaged that will host new facilities for the community, such as a health center and a community center (Figure 11.2).

The proposed *regeneration* practice will change the destiny of a place that was neglected and abandoned, while after the requalification, many new activities will be promoted such as sport, cycling, and walking. The *technological approach to*

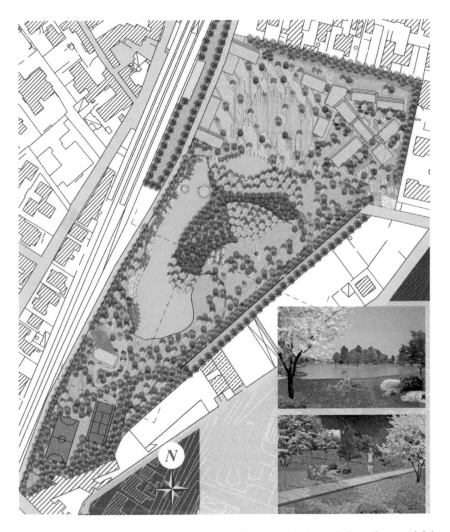

Figure 11.2 The disused area in Casalnuovo: project general plan and views (bottom right)

the environmental design is reflected in the selection of materials, most of which are at zero km availability, the use of soft technologies, and the opportunity of reusing some old parts of the existing area, in doing so, reducing the ecological footprint. For the reuse of this area, some transportation improvements are needed; therefore, engineers and planners will form part of the *expertise* involved in the completion of the project, together with the architects, hospital experts, and landscape specialists.

c The requalification of the hemp factory in Frattamaggiore

In the area of this project, an unused old hemp production factory has been abandoned and neglected for many years and the Town Hall held a competition for designing new systems and proposing new activities for the abandoned area.

The main goal of the proposal[3] was using the potential of hemp as an innovative material, which is sustainable and important for cultivation. Wherever hemp is growing, a lot of pollution can be absorbed, so allowing the soil to become healthy for the population who live in its territories. The whole area, a triangular shape on the plan, which contains some long buildings that once hosted the hemp factory, can be requalified by recovering not only the existing buildings—so saving the materials and the resources from being demolished and also reconstructed—but also the opportunity of showing to the population the importance of this product as an innovative material that can be employed in construction. By requalifying the existing buildings, some applications have been proposed and designed for the hemp itself as a construction material, which has economic appeal in the innovation of the green market sector (Figure 11.3).

The *technological approach to environmental design* is mainly demonstrated by the selection of those materials that are mainly existing and local, such as the hemp itself, timber, and rammed earth. As far as the *regeneration* is concerned, the intended use is as a museum for both traditional hemp production systems and other activities aimed at teaching the new generation how the hemp itself can be reworked and handled. This can help in making new products that reduce the ecological footprint, not being either artificial or oil-derived, while enhancing the local economy. The *scientific analysis* was aimed at investigating the solar position towards the buildings to improve the bioclimatic position and comfort, as well as collecting the solar radiation to reduce fossil-fuel consumption.

The local material culture, which was clear in the existing fabric of the factory, will be conserved and enhanced to show clearly the historic building technologies such as vaulted ceilings, thick walls, and masonry. The *expertise* required for this project are the producers of the old and traditional methods for using hemp and arts and crafts, who could still teach the new generation the secrets of their traditions. In addition, planners, urban transportation engineers, material experts, historians, and restoration specialists for analyzing and conserving the existing structures would be required.

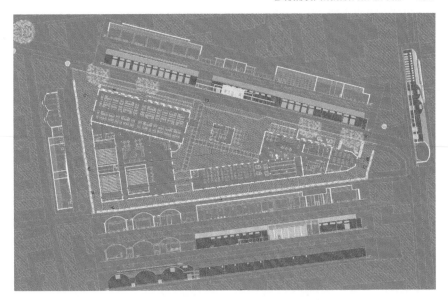

Figure 11.3 The old hemp factory: requalification plan and sections

d The Saglikli Yasam Park in Istanbul

The issue of immigration into the urban and peri-urban areas of the metropolis can be faced from a number of viewpoints. The environmental design approach focuses on the possible modifications due to the social changes occurring within the territories and the potential for developing sustainable policies of regeneration. The way in which the immigrants manage to organize their lives, in terms of work, habitation, interpersonal relationships, etc., in the host country and in the specific locality does transform, inevitably, the image of the city. Global or *globalizing* cities (Sassen, 1991) are attracting various categories of migrants; hence, they increasingly become multicultural entities characterized by pluralism and diversity (Papastergiadis, 2000). Migrants interact with the local population, which gradually results in the breakdown of cultural barriers and prejudices. Spaces of interaction include the neighborhood, school, etc., but the workplace appears to be the primary sphere where migrants and locals have the chance to meet, cooperate, talk, and socialize. Such migration patterns within towns and peri-urban areas also raise various sustainable development concerns in the growing cities such as employment, housing, health, education, and infrastructural needs and demands. This leads to issues related to environmental concerns, social hierarchies, segregation, communal disputes, and other development concerns.

The project of Saglikli Yasam Park in Istanbul aims to requalify a peri-urban zone of the city with temporary facilities for immigrants, through provision of a small and low-rise hosting center for the immigrants themselves. Three large green areas are included, besides a market and a central green plaza connected

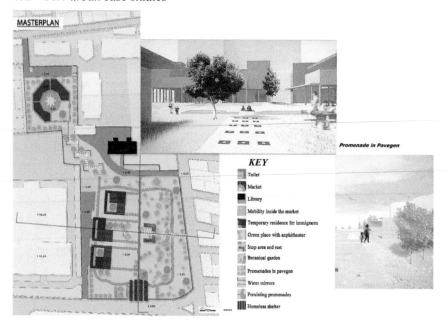

Figure 11.4 Sagliky Park in Istanbul: masterplan of the designed proposal and view of the path

to the market by means of a cultural path (Figure 11.4). The *city regeneration* is aimed at improving the usability of the park, putting together the immigrants and the autochthonous population and creating spaces of integration.[4] The *technological approach to environmental design* is achieved because the only employed technologies are soft and sustainable and can actually improve the quality without increasing the impacts. In particular, Pavegen technology was used for the path between the market and the plaza, which generates piezoelectric energy from people's footsteps when a pedestrian walks on the paving. Many *experts* are needed such as botanists, engineers, planners, architects, and sociologists to address the immigration questions. A number of *investigations*, such as solar radiation, natural light infiltration, the planning situation, and the green presence, were necessary to inform the project decisions. These are the height of the immigration hostel, where the natural light can enter, and the other solar and wind devices used for the paths' protection.

Notes

1 A. De Risi, degree thesis in Architecture, University of Naples, "Federico II."
2 P. Errico, degree thesis in Architecture, University of Naples, "Federico II."
3 L. dell'Omo, degree thesis in Architecture, University of Naples, "Federico II."
4 R.S. Galletta, degree thesis in Architecture, tutored by Profs. M. Clemente and D. Francese, University of Naples, "Federico II."

12 Hyper-commercial areas

Introduction

Commercial areas have recently been identified with a specific architectural style, that of the hyper-places. Usually, they can be found on the outskirts of European cities and they are becoming part of the already described changing territories.

> Such transformations which take place outside the urban cores can be summarised by the term peri-urbanisation. However, with this very broad definition, peri-urbanisation overlaps and coincides with many other phenomena and dynamics elaborated and described by researchers in the last decades. Besides commercial and infrastructural development, the internal migration pattern represents a major driver for peri-urbanisation.
>
> (Zasada *et al.*, 2011: 60)

The term, commercial, comes from the ancient Latin, *com*, which means with, and *merci*, which means goods for buying and selling. In ancient times, commerce was a kind of holistic exchange, not only aimed at, precisely, buying and selling, but also at meeting, enjoying things, discussing, talking, playing, and exchanging opinions: a little bit like the *Forum* in ancient Greek and Roman towns. Now, again, in these hyper-commercial areas this role can be seen:

> [In fact] the socio-anthropic change…has determined the passage of the shopping malls from places of consumption to places of socialization, expression and communication, which tend to substitute the usual locations aimed at these functions: bars, coffee shops, squares, dancing areas, public gardens… The shopping centres are places to go wandering and walking, to go and get an ice-cream on Sundays, a pizza, some shopping, to select from different shops, to go to meet people…a lot of these centres are provided with cinema halls, with baby-parking, with TV mega screens and others…They present themselves as places privileged for consumption…but also places of texturing interpersonal relationships, where a ludic dimension is blowing…The consumer becomes also a social actor which operate in the consumption and is a carrier of heterogeneous and diversified cultural needs.
>
> (Sassoli, 2008: 17)

In the new vision of a shopping center, "the action of buying a good occurs within a social relationship, which contributes to provide a sense to the purchase, and often interacts…with the symbolic content of the good itself which is bought" (Sassoli, 2008: 24).

What should happen is that "more than consumption places, the shopping malls were employed as social sites" (Sassoli, 2008: 72). The products, besides being commercialized objects of corporate profit, express relationships and links, become identity languages, and expressions of choices that are completed at an existential and political level. Nevertheless these areas present many hazards in terms of energy consumption and of antisocial behavior. There are authors who also underline the negative effects of such areas:

> The places which more represent today the thought of the democratic city are the consumption sites, the hyper-markets, the shopping centres where the access to the *under-culture* of buying is favoured in any way also by re-constructing within them the stereotyped simulacra of the town: squares, streets, bar, and so on; but also facilities such as theatres, events, music, culture in fact!…The good Volcanoes do not exist.[1] The dominant culture of the consumption wisely builds illusory visions of comfort but defines inevitably the over-growth of cities and the distribution of land.
>
> (Allen, 2008)

While architectural culture actually tries to recover the style, James Wines states: "real culture doesn't happen in art galleries or opera houses, but in supermarkets and car parks – the very landscape and environments where we inhabit" (Jacob, 2009: 25).

a The flower market in San Pietro a Patierno

The area selected for this case study has the peculiarity of addressing a very common question: that of the marginal areas in between one municipality's action and another local authority's administration. It is fundamental for the modern city to improve and regenerate these border zones, which are usually neglected, but, on the contrary, often stand as an attractive territory for new metropolitan regions. In fact, providing livability and quality in these very areas where the built heritage looks less appealing, less safeguarded, and very often decayed, could enhance all the nearby municipalities. In these peri-urban zones, it can be difficult and challenging to propose traditional use types; however, they can represent a reason for real regeneration and an element of distinctiveness as well as social and environmental sustainability.

The project presented here[2] is aimed at the requalification of an area in the municipal district of San Pietro a Patierno, near Naples, which has been the object of many different proposals, but remained inactive, abandoned, and neglected. The proposed design for the area is a new flower market. In the district and in its boundaries there are many flower growth industries, which actually need a closer location for selling their products (Figure 12.1). The design is not very polluting and it does not require a great deal of construction. Some flexible and dismantled greenhouses were designed together with the required parking areas, some offices, and a picnic point. Only one small building is proposed to be built, which will be a shoe market, as it is a flourishing local activity. A vegetable park for the district is drawn with different thematic areas, while all the functions are interlinked with each other.

The idea of unifying the main commercial activities in one large area can be an element of *regeneration* for these boundary areas, given that the new green space is attractive and the access ways are located near the main traffic arteries. The *technological approach to environmental design* has been applied in a number of choices such as the employment of an ecological floor for both the vehicles and pedestrians, made up in an organic biological polymer. In addition, further decisions were the integration of the photovoltaic hanging for the car parking, the water distributor (very important in a flower market for the required continuous

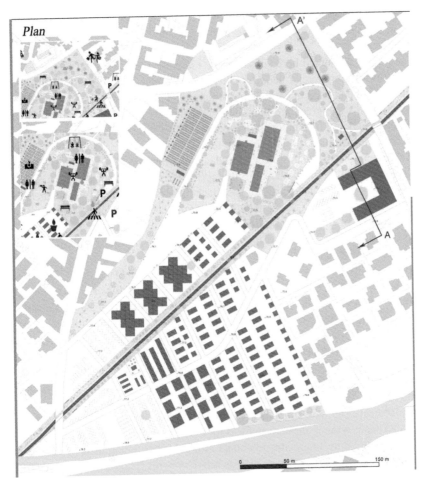

Figure 12.1 San Pietro project: general plan and detailed services (top left)

watering) made up with zero km materials, and the use of recycled PVC. Some *experts*, beside the architect, will be useful for the completion of this proposal, such as transportation engineers for the connection between the public railway and the site, planners for the legislation and the urban settlement, and some botanists and landscape specialists for the distribution of the correct plant varieties in the park. The *scientific investigations*, such as the solar radiation path, wind direction, frequency and power, sound perception, and quality of air, had confirmed the need for protecting the area from the northeast winds in winter. Furthermore, the area opens towards the south during the summer season when some sea breeze can still reach the site.

b The conversion of an old factory into a shopping center in Ercolano

Another sustainable strategy for a shopping center could involve the reuse of some existing fabrics, which can reduce the amount of new resources required for the construction, as well as contributing to the regeneration of city life in the surrounding area.

The project presented here considers the requalification of an old typographic factory into a commercial activity, where some shops, a restaurant, and other facilities can be housed in the existing structure.[4] This commercial center will be provided with some green areas, mainly in the already present courtyards, which could be turned into fruit gardens. The latter can be useful, on one hand, for proving good microclimatic conditions due to the presence of vegetation that could cool the summer days and protect from the cold winter periods, and on the other hand, for exploiting and enhancing the agricultural vocation and tradition that represented the cultural identity of the place. The existing factory is located in an extremely favorable and strategic area, very near to the—almost finished— motorway exit of the connection between Ercolano itself and the close town of Portici (Figure 12.2). Therefore, the *regeneration* of the area surrounding the new shopping center could allow an increase in the attraction of the area towards the nearby municipalities. In addition to the previously mentioned use of vegetation in the courtyards, some other devices for the *technological approach to environmental design* have also been applied. These are rainwater collection and reuse and a photovoltaic glazed surface for the areas of the building envelope with a southern exposure to employ renewables and satisfy energy requirement of over 20 percent, according to the European prescription for 2020. Furthermore, there is a green wall for the summer days, a trigeneration system, a roof garden, and a little aeolic tower for collecting the wind. The architects, plant technicians and transportation engineers will be the main *experts* that could work together for the completion of this project. The new shopping center, designed according to the *bioclimatic approach* needs a lot of climatic analysis to be sure about the solar path during the entire year, and thus, being able to collect as much solar radiation as possible for both the passive (glazed walls) as well as active systems (photovoltaic). The wind has also been *scientifically investigated* for the design of the wind tower.

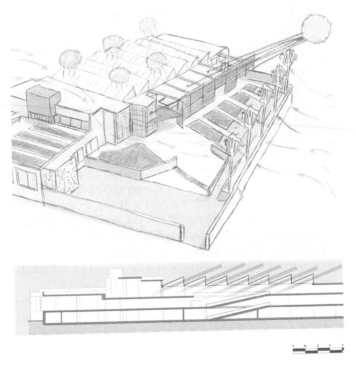

Figure 12.2 Commercial center in Ercolano: project sketches, with solar access

c The requalification of the Canzanella market in Naples

The area of the so-called Canzanella market is located in a peri-urban area of Naples, where the city has grown recently and residential settlements have increased in number and size. These spaces have only recently been recovered and destined to host a district market, which has over the years become a multigoods commercial center, not only for the inhabitants of the zone but also for customers from other parts of the city. The potential of the area is great, as, in addition to the long and articulated fabric that hosts the covered shopping area, run in the morning, there are also other activities, such as the open-air market, a little church, some green areas, and a parking zone. The whole area is fenced with a long barrier, which is closed in the afternoon and guarded overnight, so preventing use by the inhabitants. The other worries found during investigation studies were the abandoned state of the little green park, the inefficiency of the car parking, and the disorder in the open-air market area (which is now freely accessible and left in a very dirty and untidy state after closure). Further concerns were the roof of the covered shopping area (which does not protect from the rain in winter and becomes overheated during summer) and the problematic car and lorry circulation.

The proposed project[3] was aimed at reconnecting all of the area with the external residential spaces by eliminating the fence surrounding it and increasing the number of access points and entrances now closed or disabled. The main existing entrance on Metastasio Street will be doubled in size to separate the pedestrians from the vehicles. The entrance from Via Bixio will be reopened and changed to differentiate the cars incoming from their outgoing and from the pedestrian access. The internal areas have also been rearranged to satisfy this goal: the pedestrian and the vehicular routes are thought as completely separate, as well as the parking areas, which will be located where they are needed. Thus, the entrance for the lorries transporting the goods to be sold is near the two market areas, the car parking for the shopping customers is near the two main entrances, and the bicycle and motorcycle parking is near the small entrance. A long pedestrian road, connecting the upper part (which presents another entrance with two elevators) with the entrance on Via Metastasio, represents the main line that governs the redesign for regenerating the place, without constructing new fabric or involving the use of raw materials; therefore, saving precious resources.

On the east and west sides of this road, the project is developed by reorganizing the various activities. On the east side, the open-air market will be drawn into a circled area, with the center in the existing little church to rationalize the selling activities as well as to allow the customers to interact, rest and to feel free, with the presence of some sitting areas and some shading vegetation. On the west side, the green area will be redesigned to avoid the abandonment question. This zone will include other activities such as field sports, children's playgrounds, and a bar (which will be located in the existing fabric, now neglected and which can be rehabilitated); therefore, the area could be valorized and maintained in a better state. The central part of this circle will be aimed at hosting an open-air theatre to promote the use of the Canzanella zone during the evening (Figure 12.3).

The *technological approach to environmental design* is demonstrated by the use of local stone for the new design of the floors in the pedestrian areas and in the decision to organize an area intended for the didactic activity of the recycling issue. Further approaches are the sensorial and urban fruit gardens, the use of recycled materials, the new sitting and resting areas, the employment of photovoltaics on the roof of the covered market, and the use of a new system in the *rain garden* for collecting and recycling the rainwater. These ideas were achieved only with a very careful preliminary analysis and *investigation*, aimed at clearly identifying the solar, wind, vegetation, and water cycles. The *experts* needed for the completion of this complex project will be the planners and transportation engineers for the rearrangement of the external roads to facilitate the new access layout and architects and energy engineers for all the bioclimatic devices. Additional experts are botanists and landscape architects who are needed for the redesign of the small park. Furthermore, in the resting areas, they will select evergreen species for wind protection and deciduous vegetation for shading the sun in the summer and collecting it in winter.

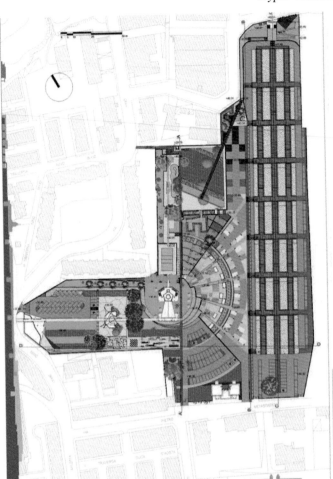

Figure 12.3 Canzanella market regeneration: general plan and sections

Notes

1 This term "Good Volcano" refers to a big shopping mall, designed by Renzo Piano, which is near Naples (Nola) and is in fact called good volcano, in analogy with the actual volcano (bad for causing disasters) that overlooks Naples, Vesuvius.
2 C. Piccoli, degree thesis in Architecture, co-tutored by Profs. M. Russo and D. Francese, University of Naples, "Federico II."
3 M. G. De Riggi and R. De Girolamo, tutored by Profs. D. Francese and A.M. Puleo, University of Naples, "Federico II."
4 D. Amodeo and C. Ciccarelli, designed for the thesis in Science of Architecture, University of Naples, "Federico II."

13 Wetlands

Introduction

The wetlands, being very humid, are part of a very important bioregion, which needs to be protected as well as enhanced to teach young people the importance of ecosystems and to save protected species of animals and plants. Over the years, the wetlands have assumed the meaning of a symbolic indicator of nonpollution, health, and biodiversity, as the water and the damp landscape is, as it is natural, a source of life for many biotic organisms. Usually, sites with such characteristics play an international role as habitats for water birds; therefore, many scientific communities and experts have finally agreed on common intentions. According to the International Ramsar Convention, the wetlands include:

> All lakes and rivers, underground aquifers, swamps and marshes, wet grasslands, peat lands, oases, estuaries, deltas and tidal flats, mangroves and other coastal areas, coral reefs, and all human-made sites such as fish ponds, rice paddies, reservoirs and salt pans.[1]

The wetlands are, of course, precious for the human cultural and scientific heritage, and any intervention in these areas should be well calibrated and should respect the existing species to as great an extent as possible, if not even promote their eventual repopulation. One of the three fundamental goals of the Ramsar Convention is that any Contracting Party is committed to "work towards the wise use of all their wetlands."[2] Therefore, the design procedure for regenerating such areas should be more cautious than the others, and it is precisely here that the bioregionalist approach, the ecosustainable technologies, and the biocompatible materials and products find their actual purpose.

a The requalification of the Sant'Ambrogio site

One of these wetlands has been the object of requalification.[3] The selected site is located in a rural area, not far from the city of Salerno in southern Italy, in the Municipality of Montecorvino Rovella. The place presents a small river, called the

Renna, which runs all along a great growing plantation of Italian nuts, and a little, very ancient, chapel dedicated to Sant'Ambrogio, southern bastion of the Long-beard's domination in Italy, from around the medieval age.

The goal of this requalification was mainly that of enhancing the entire abandoned area, by valorizing not only the zone around the chapel itself, now decayed and uncomfortable, but also the surrounding country where the nut plants could be enjoyed with some cultural as well as ecosystem tourism. The humidity around the area, due to the presence of the river, could represent an obstacle to comfort during winter, while it is a real resource during summer. Some proposals have been made according to these needs, all oriented towards a restoration of the *parvis* of the small church of Sant'Ambrogio, as well as towards a re-employment of the surrounding area.

The small paved zone near the church is completely deprived of any kind of communal facilities, if one excludes the small iron bench, located in a very uncomfortable place, and a street lamp that is broken most of the time. The access to the area should also be enhanced, for the small road bounding the site is almost invisible from the main provincial street and there is neither a real door, nor an indication, visual or virtual, at the entrance. The requirements that the project should address were the following:

- the promotion of the already restored chapel of Sant'Ambrogio
- the creation of a nice milieu around the church that could be used by groups for cultural tourism
- a path to potentially include facilities and services
- the access to be visible and to be made usable and recognizable
- some additional systems for sitting and resting
- improvement of the artificial lighting for plants
- a museum and shops
- some places for meeting and resting
- the re-drawing of the surrounding open spaces, which could safeguard the hazelnut plantation
- an ecological development of the wetlands
- a potential festival of the hazelnut
- a memorial to the Long-beards at the chapel outpost.

The first proposal[4] was aimed at the requalification of the surrounding zone by establishing a common green area, in which a small bar, information point, and some other activities were considered. The small constructions, for both the bar and information point, were all conceived in rammed earth, as in the area, even though no traditional construction in such innovative and sustainable material is present, soil with abundant clay can be found, the evidence of which is shown by the great number of furnaces for the traditional ceramic arts and craft. The *regeneration* of this peri-urban area is aimed at restoring the correct historic and artistic value to the Chapel of Sant'Ambrogio, in addition to the exploitation and valorization of the naturalistic preciousness of the place. The *scientific investigations* required for the design process were aimed mainly at including the site in the ecosystem of the wetlands, so the solar radiation, vegetation, water basins, and wind direction could all guarantee the actual knowledge of the hazards and the potential of the site itself. The use of rammed earth for the completion of the small constructions, besides being important for providing the local artisans with a renewed knowledge of this technique, will also demonstrate the *technological approach* of the selected decisions. The *expertise required* for the project is made up with planners and transportation engineers for re-establishing the importance of the localization of the site, architects for the design of the rearrangement, landscape specialists for the layout of the chapel surroundings, and botanists and agronomists for the management of the nut-cultivated area that is close to the river (Figure 13.1(A)).

Another proposal[5] was mainly orientated to *regenerate* the site by providing tourists and other visitors with a refreshment area and car parking. The

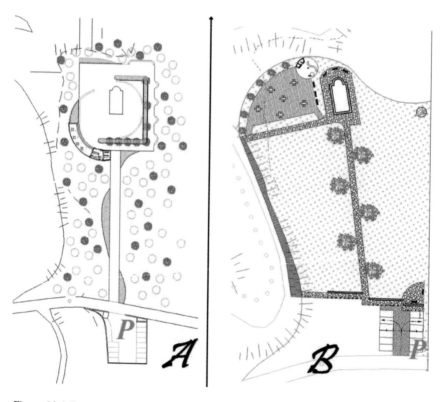

Figure 13.1 Proposals for the Sant'Ambrogio's site: general plans for (A) the first project on the left; (B) the second hypothesis on the right (on the top side: the area for lunch)

observation of the critical spots of the zone showed the absence of car parks, resting zones, lack of exploitation of the natural resources, and the bad maintenance of the green areas. Therefore, some de-bushing operation is proposed, the creation of an area for lunch, picnics, and drinking, and the insertion of a small bridge between the green area and the parvis of the chapel. The small construction intended to host the previously mentioned activities is all built in rammed earth with the technique of the *Torchis*, which is a mixed structure with timber and rammed earth. A plastic dough of earth and straw is then employed for finishing the grid made up with bamboo, canes, timber, or branches from the existing trees of nuts and Salix after having fixed it to the structural system: all choices aimed at achieving both bioregionalism and a *technological approach to environmental design*. Some structural and geotechnical engineers are needed as *experts* for the completion of this project, in addition to some architects, planners for the urban implications, and botanists. The *scientific investigations* were aimed at orientating the small constructions towards the south to be comfortable during the winter but also have good shade from the sun during summer. The humidity levels, wind prevalence, and vegetation studies were also important (Figure 13.1(B)).

The final proposal[6] shown here is aimed at enhancing the panoramic and naturalistic value of the site by *regenerating* the whole area, with the insertion of new green spots, which surround an open-air amphitheater (Figure 13.2).

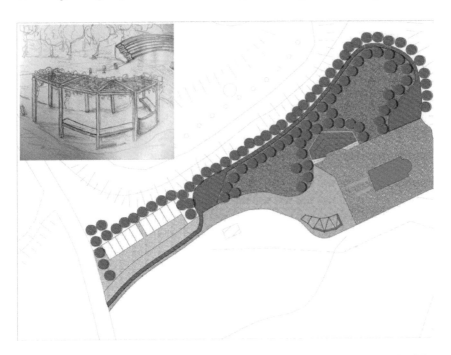

Figure 13.2 The third project for the Sant'Ambrogio site: general plan and sketch of the theater (top left)

The proposal has been drawn according to the existing panoramic sites that are defined by the existing ruins on the site. Furthermore, another zone is previewed for allowing the visitors to rest and enjoy the view and the nature, designed with a small pergola all in timber. There is also the chance of including a cycle lane that surrounds and encloses the whole site, which will be allowed to reach the new car park located near the entrance. Both the amphitheater and the structure for the pergola will be in rammed earth and other zero km products, thus demonstrating the will of responding to an *environmental design* that can be sustainable and bioregionalist. As for the other proposals, the *experts* needed are planners, architects, geotechnical and structural engineers, and landscape specialists. The *scientific investigations* were aimed at a good orientation for the theater and pergola and selecting deciduous trees, for shade in the summer and exploiting the heat of the sun in winter.

b The reclamation of the Sant'Antonio river mouth

The peri-urban area of the municipality of Sant'Antonio Abate in the metropolitan city of Naples has been neglected for a long time. It has not been clearly intended for specific activities, giving birth to a number of natural and anthropic decay phenomena. In one of these zones, a wetland is located where the mouth of a small river occasionally emerges, while it is buried at other times. The requalification and the reclamation of this area has been the object of a collaboration between the local authority and the University of Naples. A brief was defined for the intended use and exploitation of the potential of the area, which is very long and articulated, and provided with cultivated, as well as still naturally conserved, ecosystems.

One proposal[7] is aimed at *regenerating* this zone with new structures and services, while at the same time trying to avoid new constructions and destroying the local flora and fauna (Figure 13.3). Within the boundaries of the abandoned area, some new activities will be included for valorizing the natural and the few anthropic presences. These activities will be mainly aimed at bird watching, photography, fishing, and arranging a natural swimming pool and picnic area. Some zones are proposed that are intended to host scholastic trips with an interactive program thanks to the inclusion of an urban fruit garden. A small flower market is considered in some areas that are close to the cultivation of the flowers themselves and some other didactic and sport activities in the open air. All the small new structures that will be built are conceived in timber and the inclusion of a phytodepuration area demonstrates the *technological approach to environmental design*. Some *experts* will be needed for the completion of this idea, such as natural scientists for the repopulation of flora and fauna within the site, architects and landscape architects for the redesign of the anthropic as well as naturalistic zones, and some planners, who will help to study and focus on the current local legal regulations. Other *scientific investigations*, such as the vegetation, solar access, and wind and humidity levels, will be useful for the correct environmental design.

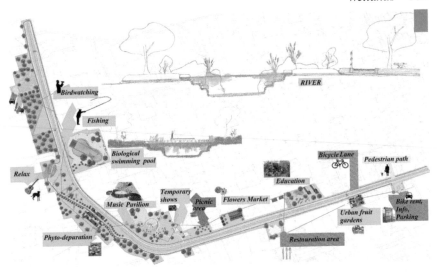

Figure 13.3 The Sant'Antonio river park: general plan of the requalification first project and sketches of the solutions

Another proposal[8] will try to regenerate and make healthy the river area by enhancing it with new systems to give this part of the land back to the town's citizens (Figure 13.4). At the northwest side of the long river shore, there will be an area destined for bicycle hire, with some advertisements illustrating the river's park and its facilities. There will also be a carpark for the people who come to visit the park and they will continue along the path on foot or riding bicycles. Some of the fruit gardens (4,800 sq. m), now owned by the public municipality, can be managed by the central town private inhabitants, and some others (2,900 sq. m) by the school for didactic goals. Continuing along the river's shore, another wide space is proposed for a structure, to be used for shows and events, in the shape of an amphitheater, made up of terraces and other ground sustaining systems according to the naturalist engineering approach. A small lake will be filled in to create space for hosting various species of water flora, and will be very close to a green oasis, in which a massive increase of wood planting and an installation of structures for resting and stopping are proposed. Following the river bends, some toilets and dog walking areas will be connected to the left shore by means of the recovery of some of the existing bridges, considered with local stone and timber; therefore, answering questions in the *technological approach to environmental design*. All along the sides of the river will be the arrangement of a gymnastic path, which will be 2,400 m long. At the other end of the emerging river, on the southeast, another car park will host the visitors coming from the nearby village and some boards will describe all the paths (cycling, gymnastic, and pedestrian), together with an information point and the bicycle rental office.

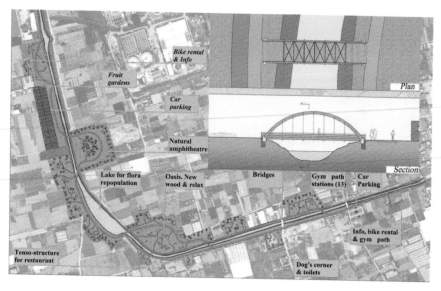

Figure 13.4 The Sant'Antonio river park: general plan of the second hypothesis and detail of the technical solution for the proposed bridge

Some planners and transportation engineers will be needed for the reorganization of the surrounding roads and paths to incentivize the use of this river zone. Other *experts* such as naturalistic engineers, landscape architects, and botanists will be required, together with some natural scientists for the repopulation of flora and fauna. The *scientific investigations* were aimed at understanding the existing vegetation and bird presence, the sun paths, and the humidity and temperature variation through the whole year, together with some acoustic analysis to locate the more fragile species in a zone which will be far from vehicular noises.

The last proposal for the same river shore[9] proposes the rearrangement of the whole area, by including two main activities, flower cultivation and repopulation of birds, with related bird watching action by visitors (Figure 13.5). Different parts will be realized completely in bamboo and timber, or in local stone, so responding to the *technological approach to environmental design*, while it will be possible to enjoy the whole area by means of the rational lines of the paths. The park will also host some new species of birds, so the park will become the location for bird watching of those species, which will be attracted by a better-chosen vegetation, with the help of some natural science and botany scientists, as *experts* of the local fauna and flora.

In fact, from northwest, the project will propose some didactic farms, a flower productive system with an exposition zone, then an hydrophilic wood, some fields for sustainable agriculture, a reading area, and a wide bird-watching space, while at the extreme southeast side, the other entrance will host an information point. The *regeneration* will then also promote some activities for children as well as adults, both locals and tourists, so valorizing the district. The small constructions for the bird-watching activities will be made up with timber, local canes, and recycled materials (Figure 13.6). They will employ technologies that could be dismantled easily and remounted elsewhere, responding to the sustainable concept of *zero-waste design*, promoted by Paul Connett (see Part I, Chapter 3). All the choices were made with a careful accordance with the local climate and other

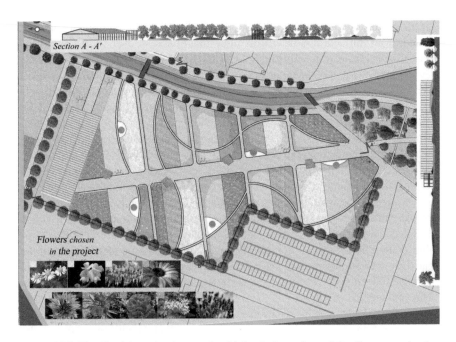

Figure 13.5 The Sant'Antonio river park, third solution: plan of the flower productive system with sections and detail of the flowers previewed for growing

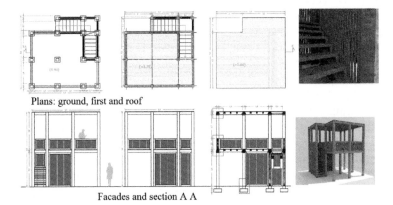

Figure 13.6 The Sant'Antonio river park: bird watching tower in timber and canes: details

environmental characteristics of the area, which had been studied with some *scientific investigations*, aimed mainly at selecting the right location for plant life as well as for the visitors' comfort.

c The requalification of the wet area in Prepezzano (Giffoni Sei Casali)

A very picturesque and historic municipality near Salerno in southern Italy is located in an intricate and beautiful valley that is spread over six different districts (Casali), from which Giffoni Sei Casali (six districts) is derived. One of the districts presents a very wide surface, and in the peri-urban area, there is a small river, called Prepezzano, from which the district itself takes its name. Here, a very attractive area close to a small mountain has been neglected and abandoned for some time. A design competition was launched a few years ago and some proposals have been made. Among them, one can be considered wholly belonging to the aims of this book, being in harmony with both bioregionalism and sustainable environmental design. Some problems have been tackled, prior to commencing the project, for example, the access: in fact, given the great size of the area and the presence of the river, pedestrians cannot easily reach and enter the area. The internal paths are not rational or adequate; an absence of attractive activities can be noticed, together with insufficient car parking spaces. There is also a large roller skating field, which is completely neglected and unused for most part of the year. The vegetation, representing a great potential of resource and richness, is in an unmaintained state. The solar radiation and the planning tools have been part of the applied *investigations*.

Following the knowledge of the site, a preliminary project[10] has been made for the regeneration of the whole area, aiming at providing the citizens of the nearby municipality of Giffoni and, in particular, of Prepezzano and Sieti, the closest districts, with a meeting point adapted to the inhabitants' needs and requirements (Figure 13.7). The physical, natural and anthropic existing lines, such as the river, the main road of the local district and the provincial street, will become the geometrical strength of the whole design enhanced by the creation of some belvederes and meeting points. The number of entrances and exits for pedestrians will be increased. A bicycle lane will run along the river, crossing the whole area. Some timber bridges are proposed that will allow both cyclists and pedestrians to cross the river in various spots: an action that is now impossible. The parking numbers will be increased to 65 units and the roller skating field will be recovered, with the chance of transforming the whole field into a temporary market for trade shows, local fairs, exhibitions, and other events. Finally, the green area is proposed to be as least anthropic as possible. The northern zone will be a green area with some facilities, comfortable paths, and services for all. In the southern zone, instead, the green areas will be left wilder, where the nuts and olive trees, now present and not autochthone, will leave the place to the Mediterranean scrub, and some of the paths will be adapted to trekking. The *technological approach to environmental design* is also evident in the decision to enclose the park borders with a vegetation fence, which is able to reduce the noise from the town that has been found during the *scientific investigation* stage to be very annoying; therefore, making the park a quiet sanctuary.

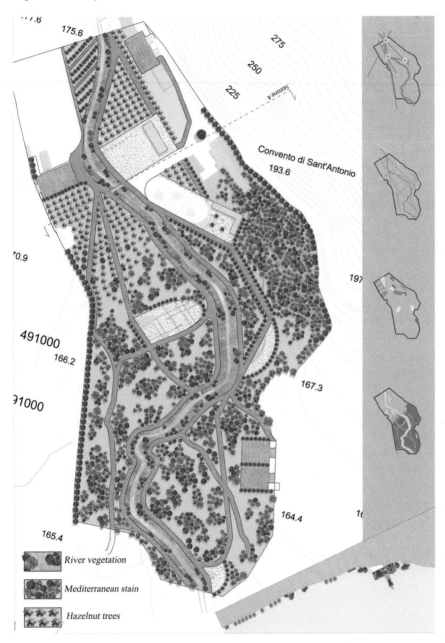

Figure 13.7 The Prepezzano project: general plan of the area, conceptual map (left) and transversal section (bottom right)

Notes

1 The Ramsar Convention mission is identified with "the conservation and wise use of all wetlands through local and national actions and international cooperation, as a contribution towards achieving sustainable development throughout the world" (www.ramsar.org; accessed 31 October 2015).
2 The other two so-called "pillars" of the Convention, established that the contracting parties commit to "designate suitable wetlands for the list of Wetlands of International Importance (the 'Ramsar List') and ensure their effective management; and cooperate internationally on trans-boundary wetlands, shared wetland systems and shared species" (www.ramsar.org).
3 The proposals shown here are the fruit of a collaboration between the University of Naples, "Federico II," the local municipality authority, and the Falcone furnace, which hosted and supported the final workshop and organized the final award ceremony.
4 The solution was conceived by the students A. Coppola and S. Aiello.
5 Designed by the participants to the workshop: D'ambrosio, Del visco, and di Mauro.
6 Solution proposed by the participants to the workshop: G. Palmari, L. Miccoli, and L. Mancini.
7 A. Amoroso and M. Casadio, tutored by Profs. A. Vuolo and D. Francese.
8 M. De Felice and D. Turco, tutored by Profs. A. Vuolo and D. Francese.
9 This solution was designed by I. La Frazia and O. Monfreda, tutored by Profs. D. Francese and A. Vuolo, as well as G. Longobardi, 2014.
10 Designed by S. Lino, Department of Architecture, University of Naples, "Federico II."

14 Peri-urban parks

Introduction

The question for the parks that are located not in the center of the city but on its surroundings can be faced from two different points of view. On one hand, these green lungs have a great importance for some areas, which are often polluted due to industrialization. On the other hand, being in peri-urban zones, they are sometimes neglected and not cared for enough, thus creating those phenomena of anthropic wilderness, changing their original agricultural vocation into a very dirty and abandoned state.

> Physical conversion of open space – in particular agricultural land – for urban purposes and socio-cultural transitions in rural areas through adoption of urban life styles or in-migration of urban dwellers, leads to the establishment of a peri-urban space, and sets different forms of urban and rural living and working into close contact.
>
> (Zasada *et al.*, 2011: 59)

In fact, the peri-urban green areas can be thought of as the following:

> [They can be] considered…as a third-ality. It is not the country space interclosed between the urbanized sites, it is not made up as an urban vegetable garden, not only inhabitants' dispersion, neither city nor country, but it is more city and more country at the same time.
>
> (Mumford, 1954)

a The park of the Villa Medicea (Ottaviano)

An ancient villa near the city of Naples, in a marginal area of the municipality of Ottaviano, has been recently confiscated by the Organized Criminality Board and destined to host the offices of the Regional Park Authority of the Vesuvius district. The Villa Medicea is located close to one of the accessible areas of the volcano itself, where a splendid panorama and fertile vegetation can be enjoyed. The projects described here were part of an exchange program between the university and the Vesuvius Authority, with the goal of improving the external area of the villa, which has some potential usage and cultural interest, both anthropic and natural, but at the moment, it is still a little neglected. The projects produced were then shown in an exhibition held in the courtyard of the villa itself.

One of the proposals,[1] aimed at promoting the little park as a didactic green area, had the idea of using the green area as lungs for the metropolitan area of Naples, which surrounds the villa and its external places and produces atmospheric and acoustic pollution, thus, *regenerating* this peri-urban zone (Figure 14.1). The real project proposes the creation of the previously mentioned didactic park by dividing the place into four different areas. The *entrance*, which will include a parking area with a timber pergola for the visitors, an information point, and a panoramic viewing spot. The *energy* area, where some installations of elements for producing energy from renewables are considered. The *biological cycles* area with some timber structures aimed at the composting process and some fruit gardens and aromatic herbs to be grown by school children. Finally, the *nature-and-sense* area, where some activities will focus the attention towards the five senses.

The future employment of natural materials and zero km products, as well as the solutions selected for the production of alternative energy can demonstrate the *technological approach to environmental design*. A number of *experts*, such as planners, landscape specialists, geologists, botanists, geotechnical engineers, architects, energy engineers, and historians are needed for the completion of the project. A number of *scientific investigations* will be fundamental: wind and sun paths, vegetation life and evolution, historical background, and geological development of the area to orient the project towards a sustainable and environmental design.

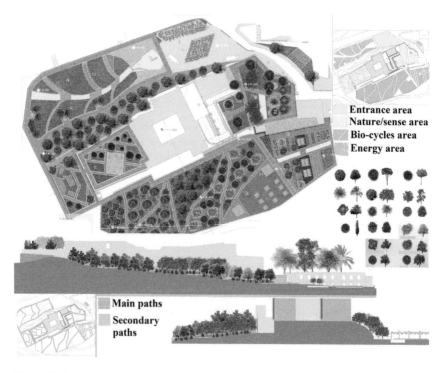

Figure 14.1 Villa Medicea project, first solution: conceptual map and functions (top right), general plan (top left), sections (bottom center and right), green essences choice (center right) and paths (bottom left)

Another *regeneration* proposal[2] for the same park is aimed at enhancing the area by including some structures that can improve the human senses, with a peculiar project (Figure 14.2). The area around the Villa Medicea, seen as a place for the renaissance of the senses, will be requalified with the insertion of some aromatic plants. They will stimulate the sense of smell and additional varieties will be used for tasting, some others will be adapted to promote the visual pleasure of strong colors (for example the *Cercis siliquastrum*, commonly known as the Judas tree) and beautiful shapes. In the zone south of the entrance, the visual sense will be stimulated by promoting all the park activities. There will be some didactic fruit gardens for taste and then a path for visual panorama. The aromatic garden will be placed on the other side of the villa, in the western zone, where the sun rays, being evaluated during the *scientific investigations*, have been found to be strongest in the afternoon. This enhances the various smells and perfumes coming from the selected trees and bushes, directed also by some *experts* in botany and landscape. On the northwest side, where the sun is often absent, the musical garden will be placed, with benefits coming from the nearby Vesuvius great park area, which will act as protection from noise, as the analysis has shown. The last area, the north zone, is considered for ludic activities for children and adults, thereby stimulating the sense of touch. Some new paths are proposed for connecting the various zones; all completed in local timber or in waste recycled materials, so matching both the sustainable *technological approach to environmental design* and the *bioregionalist* concept. In addition to the previously mentioned botanists, engineers and planners will be useful for the reorganization of the area surrounding the park, helping to gain access for pedestrians as well as easy parking for cars.

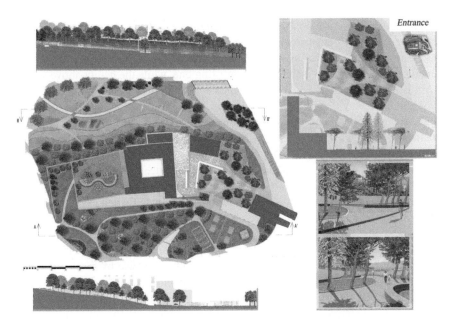

Figure 14.2 Villa Medicea project, second solution: section (top and bottom left), general plan (center left), detailed plan of the entrance (top right), and renders (bottom right)

b The regeneration of the Jardines de la Menara (Marrakech, Morocco)

Under the framework of a convention between the University of Naples, "Federico II" and the University of Marrakech, Cadi Ayyad, an experimental work has been carried out for the promotion and development of the ancient Mediterranean material of rammed earth. This material has been neglected for many years as it was considered as a poor system of construction. It has now come into popularity again and some applications have been made for its sustainable and bioregionalist potential: the Pavilion of Morocco at the Universal Expo 2015, in Milan, is built using this technique.

A neglected peri-urban area just outside the city of Marrakech is not rationally designed and the local administration had the intention of reorganizing and requalifying the entire area. Therefore, it has been the subject of a number of projects, aimed at the regeneration and valorization of the place. The area, le Jardines de la Menara (the gardens of the Menara), hosts an ancient pavilion, a water basin, and a large vegetation zone, mainly with olive and palm trees. The underground soil is organized as a system for water collection and distribution, for (as it is well-known) this African city has been founded in the desert and all the required water has been extracted from the ground. Therefore, the area has a very important meaning both in the physical sense, being the hydric reservoir for the citizens, and in the metaphoric sense, for it represents the life essence of the city itself.

The analysis carried out demonstrated the need for improving the area by enhancing its functional aspects, which are now represented by the affluence of the tourists to the green zone, the pavilion, and to the wide basin, which cools the hot Moroccan summers. There is also a staircase on the north side of the basin, which was destined to host spectators during water shows that are very seldom held. The quality and functionality of this staircase is very low and the material employed, iron, is not sustainable. The existing condition of the paths and main structures is also very decayed and often redundant. The basin zone where the small pavilion is located is the more interesting one, while the height above water level is useful for the hydric distribution for irrigating the plants.

Although some potential can be found in the geometry of the area, which has a quadrangular plan and a very orthogonal layout, the strict line connecting the main entrance to the long internal path, which also ideally continues the line of the Bab Jadid square (one of the doors of the walls of the medina) until reaching the Koutoubia, the most important and frequented Mosque, is brusquely interrupted by a large wall. This interruption leaves the visitors bewildered and disoriented, while they have to take secondary paths for reaching the basin and the small pavilion with the museum. The whole area is in a semi-abandoned state, both the gardens and paths (Figure 14.3).

One of the proposals[3] for the regeneration of the gardens is aimed at improving the fluidity and empowering the functionality of the area, also by means of some constructions in rammed earth. Resting areas and an information kiosk at the end of the long avenue will be proposed. The staircase will be eliminated and substituted by a sustainable terrace for a restaurant, aiming at enhancing the continuity of the landscape by promoting the use of the existing spaces beneath, which are already well equipped. This platform will be built in timber and rammed earth (with the technique of the *pisé*) and covered with a framed roof, also in timber from olive trees (Figure 14.4), responding to the *technological approach to environmental design* as well as to the *bioregionalist* criteria, for both the olive trees and the earth are locally found.

Figure 14.3 The park of "La Menara," in Marrakech, existing situation: old picture (top), view of the basin (middle left), the restaurant (bottom left), and the entrance (right)

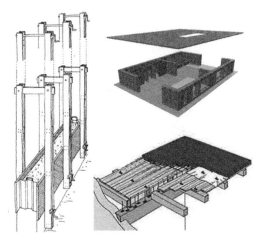

Figure 14.4 First solution for the regeneration of the park "La Menara." The proposed new platform in timber and rammed earth: axonometric section (top right), detailed drawing of the rammed earth "torchis" (left), and detailed drawing of the roof (bottom right)

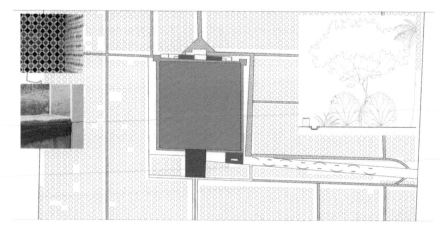

Figure 14.5 First solution for La Menara park: general plan (center), section (top right), and detail of the ceramic and raw earth bench (left)

From here, the gardens and the whole city can be viewed and enjoyed. Some free spaces among the olive trees will be provided with facilities, such as sitting zones, and the secondary lanes will be finished in trodden soil. The other main path, with the cypress trees, sitting systems, and plants that are lit-up will be rearranged. Following the *scientific investigations* regarding the shaded or not shaded places, the more humid and ventilated spots will have vegetation increased to provide protection from the summer sun and some sitting systems have been proposed and designed. The technique employed for constructing such benches is made up with *pisé* with a superior finish in ceramic, which is also a material of the local traditional artisan culture, which continues to flourish (Figure 14.5). The *experts* required for completing this project will be the hydraulic and naturalistic engineers, landscape and technological architects, materials' specialists, and botanists.

Another project[4] for the *regeneration* of the Menara gardens is instead aimed at enhancing the local traditions, which includes an atelier for the arts and craft operations in the new proposed activities. The access will be improved and little walls, passing over the level differences between the two zones, are proposed, which will be realized with the double function of providing protection from falling and of the use as benches. The paths can be reorganized by separating cars from the pedestrian flow. Some terraces will replace the existing one, which will be used for attending the events and shows occurring in the water basin and a new important function will be introduced in the gardens, which can stimulate and revitalize the whole area: an oil mill. The whole activity is concentrated in the northeast zone of the garden area, where another still squared site, more circumscribed, encloses the main artisans and local activities, the olive press (eventually fed with solar energy), and the other linked activities. The location near the road will allow lorries to arrive, load and deliver the olives and the oil, without disturbing the other activities of the gardens, which, at the same time, can become an interesting visiting spot for the tourists (Figure 14.6).

Responding to the *technological approach to environmental design*, it is proposed that the little building hosting the oil mill will be constructed completely in *pisé* and timber; it includes some vegetation and a pergola for the hot summer days. By means of a number of *investigations*, it has been possible to choose the most appropriate locations for the walls/sitting systems, the oil mill, and the other resting areas. According to the sun path, the resting zone should be sunny during winter (which is not warm, but often cold due to the wind coming from the Atlas Mountains), while fresh during summer (Figure 14.7). The wind, local vegetation, and traditions have also been analyzed, with the help of some *experts*.

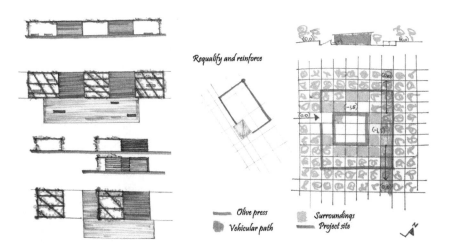

Figure 14.6 Second project for La Menara park: sketches of the design for the bench/ pergola (left), concept (center), general plan (center right), and section (top right)

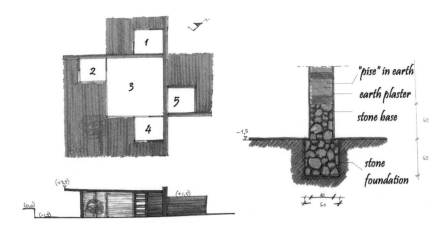

Figure 14.7 Second project for La Menara park: plan and section of the olive press (left), detail of the building wall-section (right)

c Requalification of the eco-museum of Lecreti (Benevento)

In the southern Italian country, in a very hilly and green region, some traditional constructions in rammed earth have been found, which are very old and in a decayed and neglected state of conservation. The project described here was intended to requalify not only these small constructions, but the whole area, by enhancing the rich landscaping value of the place, and promoting a kind of eco-museum[5] for the district, which could attract visitors and raise awareness and appreciation of the area. The proposal[6] was aimed at *regenerating* the district by valorizing the existing resources, both natural and anthropic. Starting from these goals, the whole site has been redesigned and divided into various zones, connected to each other by means of natural and sensorial paths. The eco-museum will contain some didactic evidence of the once active fabrication of tobacco and a space for an exhibition of the cultivation tools and modalities, together with the green areas dedicated to the actual growing of the tobacco plant itself (Figure 14.8).

The second activity to be promoted will be the knowledge and enhancement of the use of the ancient and sustainable technology of rammed earth. This other function will be developed by means of the recovery of existing fabric, which is the remains of an old rammed earth building. This building will be requalified and consolidated by means of the substitution of the almost collapsed roof and of a small extension. In this building and the pertinent areas, a zone for the explanation of the adobe technique will be located, together with a covered area for drying the earthen bricks, while the internal rooms, enclosed in rammed earth walls, will host a center for the research of rammed earth. There are two green areas: one, as previously mentioned, is destined for the growing of tobacco, the second, with the colors and odors of nature. Here, beech (*Fagus sylvatica*), olive trees, white mulberry (*Morus alba*), and other fruit plants will be increased, and some additional trees will be planted, including figs and walnuts. Some places for resting and a big terrace for the observation of the landscape of the whole eco-museum are proposed.

Other museums will be located within the existing construction after the rehabilitation, aimed at showing the history of the local municipality, the ancient town of Paduli, traditional agricultural equipment, and tobacco's history, together with a tasting room. The *technological approach to the environmental design* is shown through the selection of local and highly natural materials such as rammed earth, sandstone, and some timber structures. All the locations chosen for the trees, the bushes, and the activities in the fabrics were defined by a number of *scientific investigations* about the solar path, ventilation, humidity, noise, and quality of air. Some *experts*, such as local artisans, will be supported by architects, landscape specialists and botanists, together with transportation engineers and planners.

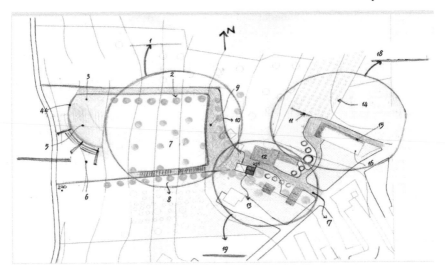

Figure 14.8 Lecreti eco-museum: sketched plan for the requalification of the area with functions (1, nature colors; 2, figs plants; 3, garden roof; 4, earth walls; 5, research center for rammed earth and laboratory; 6, covered open area for drying adobe bricks; 7, olive trees; 8, nut trees; 9, landscape and agriculture field observation points; 10, mulberry; 11, tobacco fields; 12, offices; 13, agricultural tools exhibition; 14, existing tobacco fields; 15, corridor for observing the stages of tobacco production; 16, tobacco drying; 17, new flooring similar to the historical center; 18, old historic pictures; 19, agricultural life)

Conclusion

After having outlined the description of some projects for the regeneration of the peri-urban areas, it is noticeable how the importance of the marginal areas around large cities play a fundamental role in the beauty and fruition of territories. There is presently a new trend for these neglected areas, as it has been found that they save the identity and cultural values of the population as well as their social essence, which is encapsulated in a new term:

> Paesology (town-landscape-logic) [which] is halfway between ethnology and poetry, is not a human science, is a surrendered science, useful to remain defenceless, immature…The paesology is none other than the transfer of [one's] body through the landscape and the transfer of the landscape through one's body. It is a discipline founded on the earth and the flesh. A shape of floating attention, in which the observer and the object of the observation often appear to change their role. Then it is the earth that investigates the observers' moods. The paesology is simply the writing which comes after

having wetted the body under a place's light. The paesology is [one's] way of non-surrendering to the universal up-breathing of beings and things. A shape of intimate resistance, but not for this lack of a political vein. … It is not idolatry of the local culture. …In the small town like in the big city the so-called individual's society occurs, with the autistic following democracy. From this point of view there are no differences, but it is necessary to consider that the metabolism of a place in which there are 1,000 individuals in 30 square kilometers is very different from that in which there are 30,000 individuals in 1 squared kilometer.

(Arminio, 2011: 10–12)

Notes

1 Solution designed by the students, F. Orefice and T. Milo, University of Naples, "Federico II," 2013.
2 The solution was conceived by the student, L. Nunziante, University of Naples, "Federico II."
3 Solution by M. Scavino, C. Scarpitti, and A. Romano.
4 The project was conceived by C. Arturo, D. Petrone, and S. Sposito.
5 An eco-museum is a museum that is used for telling the history of the territory and for valorizing and diffusing previous activities, such as industrial or artisanal archeology.
6 C. Recalina, thesis in Architecture, 2009, University of Naples, "Federico II."

Conclusion

The proposed methodology for investigating and designing an accessible and tangible urban regeneration, both social and cultural, has been discussed, clarified, and summarized. Possible ideas for its innovative application with the use of sustainable technologies have been suggested. This research can be concluded by underlining the importance of environmental and landscaping studies dedicated to the humanized zones, mainly when they appear strongly integrated and interacting with the biotic processes. It is important to remember that the rural landscape, which includes these two characters, actually requires extensive analysis for the understanding and adjustment of development lines. Only by these means can a project of transformation lead towards a real sustainable evolution of the areas and truly succeed in creating a valorization of the habitats, that is, a balanced share between aspects of environmental, cultural, and overall social requalification.

"The way of living the space, as well as the city-space, by the children is similar to the poet's one: he lives poetically the space, he does not submit to it passively" (AAVV, 1999: 18). Therefore, the lesson learned by children is that the city should be enjoyed in a nonalienated way, but living, here and now, any event or activity. "For citizens, for the human relationships, for the individual and social development, the quality of life depends on the value of the urban environment, either it is green or all built up" (AAVV, 1999: 19). The solution is the following:

> Opening new (physical and mental) spaces where the process is more important than the project and where, temporary, the non-built is more important than the built. The subtraction re-takes its additional character so as to imagine new relational and common goods, able to add to the community new civic virtues, new urban sensitivity, in which the belonging and the identity will not have the character of an island or an enclave, but which will define the new will of hybridizing oneself, based on the concepts of inclusion and fertility.
>
> (Persico, 2013: 11)

Hence, it will be possible to prevent the risk that "nature is being murdered by anti-nature, with humanity doing the killing, and perhaps committing suicide in the process" (Lefebvre in Neuman, 2012: 162). Finally, it can be concluded that the actual regeneration for urban open spaces with bioregionalist and sustainable technologies will employ fewer constructed elements to confirm Laozi's words:

"...who acts ruins and who keeps loses.
Therefore the wise man
Does not act and then does not ruin,
He does not keep and then does not lose.
Common men while treating the things
Often damage them at the last moment.
Pay attention at the end as well as at the beginning
And then no enterprise would fail..."

(Laozi, 1993: 143)

Glossary

Biocompatibility: 1. Capacity of materials, products, systems and fabrics to avoid damages, diseases, and/or discomfort to users. 2. The term is composed by a prefix *bio*, coming from the ancient Greek (life) and the compatibility, deriving from the Latin *cum patior*, participating to, being in harmony with, and underlines the character of harmony with the living milieu (Carbonara and Strappa, 2013). 3. The materials can be defined biocompatible when they are equilibrated with the environment and do not create any harmful effects on the health of living beings, comfort, or allergy.

Bioclimatic architecture: 1. A cultural movement that wishes a scientific control of climatic (sun, wind, humidity), edaphic, and hydrographical phenomena during the conceiving and constructing processes of human habitat to improve the internal comfort and limit the use of energy from fossil fuels. 2. The studies of the dynamic integrations between inhabiting and constructing, which will take into account, besides the climate, other elements such as water and vegetation to collaborate in the provision of comfort and energy saving (Francese, 1996).

Bioregionalism:

1. Both a geographical ground and a ground of consciousness to a place and the ideas that have developed about how to live in that place. Within a bioregion the conditions that influence life are similar and these in turn have influenced the human occupancy.

(Berg and Dasmann, 1977: 399)

2. A decentralized form of human organization, which, by aiming at maintaining the integrity of the biological process, of the life patterns and of specific geographical configurations of the bioregion, can help the material and spiritual development of the human communities they are inhabiting.

(Rebb, 2005)

3. The bioregionalism recognizes, feeds, sustains and celebrates the local links: earth, plants, animals, fountains, rivers, lakes, underground waters, oceans, air, family, friends, neighbors, communities, native traditions, indigenous systems of production and commerce.

(Berg, 1984)

Byproduct: "The residuals originated by a production process whose main goal is not their production" (Italian Law Decree, 2006). 2. "A secondary or incidental product. The result of another action, often not foreseen or intended" (*Random House Webster's Unabridged Dictionary of American English*, 2015).

Carrying capacity: 1. "The maximum population size of the species that the environment can sustain indefinitely, given the food, habitat, water, and other needs available in the environment." 2. "The environment's maximal load, which is different from the concept of population equilibrium" (see Rees and Wackernagel, 1994; Hui, 2006).

City: 1. "Town of greater importance, or size, or with wider municipal powers, than those called simply towns" (*Oxford Illustrated Dictionary*, 1982). 2. "It is a living organism, it is a complex. As such we will never be able to plan the city. But rather, at best, grasp several clues…the complex character of a city is not a problem, it is a higher gift" (Thwaites *et al.*, 2007).

Cityscape: 1. "The visual appearance of a city or urban area; a city landscape" (*Oxford Illustrated Dictionary*, 1982). 2. "The appearance of a city or urban area, especially in a picture; a picture of a city" (*Random House Webster's Unabridged Dictionary of American English*, 2015).

Comfort: 1. "The whole of conditions referred to states of the building system suitable to life, health and running the users' activities" (Italian Standard "Norma UNI/CEE 0050"). 2. "An harmonic state of health, physical and moral strength" (Devoto and Oli, 1981).

Composting: Involves collecting organic waste, such as food scraps and yard trimmings, and storing it under conditions designed to help it break down naturally. This resulting compost can then be used as a natural fertilizer.

Cultural landscape: It includes both the natural and the anthropic components and it is configured as a harmonic system between the biotic (flora and fauna) and the abiotic elements of the territory (mountain, plan, morphology, soil, rivers, lakes, and climatic conditions).

De-growth: 1. "A policy design …of constructing, in the North as well as in the South, convivial, autonomous and sober societies" (Latouche, 1999). 2. "It is not a negative growth. It should be better to talk about *a-growth* as it is talked about *a-theism*. In fact, it deals exactly with the abandon of a faith, a religion, that of finance, progress and development" (Latouche, 1999).

Desertification: 1. "The process which leads to an irreversible reduction of the soil capability of producing resources and sources" (FAO UNEP UNESCO, 1979). 2. "Decay of lands …due to a number of causes, such as the climatic change and the human activities" (UN Convention, 1977).

Ecosustainability: The entire characteristics of materials, systems, products, and fabrics, able to limit the impacts on natural ecosystems and cultural habitats.

Ecosustainable architecture: "The activity of conceiving, designing, disusing architectural works in harmony with the anthropic and natural involved systems, made up of objective and measurable factors as well as of random and perceiving elements. The ecosustainable intervention in architecture should then be compatible with the wider system in which it is included; and this in accordance with a concept of economy which is not identified as a small cost in the short term, but by considering all the issues of the question, in [a] holistic vision" (Carbonara and Strappa, 2013).

Ecosystemic facilities: "All the benefits that human beings take from the ecosystems" (Peccol, 2013).

Energy recovery from waste: The conversion of nonrecyclable waste materials into usable heat, electricity, or fuel.

Energy saving: An approach to urban regeneration that considers and analyses the peculiar phenomena that occur within a city, which can help to create comfort conditions, or on the contrary, to provide disadvantages and discomfort to users.

Landfills: Engineered areas where waste is placed into the land. Landfills usually have liner systems and other safeguards to prevent pollution to the ground water.

Landscape: 1. "It means an area, as perceived by people, whose character is the result of the action and interaction of natural and/or human factors" (Landscape European Convention, European Council, article 1). 2. The term landscape, which differentiates substantially from land and environment, has assumed with time, wide and various connotations, all strictly linked to the concept of perception. An old meaning is that of a "picture … or part of one representing inland scenery; actual piece of such scenery" (*Oxford Illustrated Dictionary*, 1982), mainly underlined on the sight, and identified with "sightseeing, panorama; part of territory which can be embraced with the glance from a determined point" (translated from *Enciclopedia Treccani*). This focuses the attention, even for the geographic disciplines, on the very sense of vision:

> the whole of elements which constitute the physiognomic features of a certain part of the earthen surface; it can be considered as the abstract synthesis of the visible landscapes, for it focuses them only on the characters that present the more frequent repetitions over a more or less big space, superior in any case to that included into a unique horizon.
>
> (translated from *Enciclopedia Treccani*)

3. The landscape is a palimpsest which saves in the present and the future, the signs of past; territory is a parchment constantly cancelled and re-written, but it leaves, without will and without knowledge, some traces [of the past territories] of which the archaeology makes its dairy.

(Raffestein, 2003: 33)

Level of naturality: "The path that reduces the impact of sample taking, consumption, emission, and dismissing of both gas and solid substances. When a product achieves, during its lifecycle, reduced laboring processes, it consumes fewer resources and produces less waste in the environment, its level of naturality is high, while in the opposite case, the artificiality level is high. The maximum level of naturality of a prime matter, that is the pure state, is called climax, or condition of maximal potential, while the opposite, when a product is provided with the maximum level of transformation, is completely artificial, has a low level of naturality" (Abrami, 1987: 119–21).

Material culture: 1. The whole of the experience coming from the ancient wisdom, referred to the construction know-how of a particular population or civilisation. 2. "Particular form or type of intellectual development of civilisation" (*Oxford Illustrated Dictionary*, 1982). 3. "The whole of the manufactured objects of a defined civilisation" (translated from *Dizionario Garzanti Della Lingua Italiana*, 1981). 4. The history of construction as a model of high urban sustainability, in which shape, technology, poetry, architecture, and performance were joined in order to sustain a structural level of comfort, and supported by the fecund, not yet lost, traces of the cultural collective memory and enriched by the fertilization of transcultural processes.

Material: 1. "What is needed for a defined use" (translated from *Dizionario Garzanti Della Lingua Italiana*, 1981). 2. "Components of the environment suitable for satisfying man's needs, identified as natural resources which state the base for the fabrication of finite elements" (translated from *Dizionario Garzanti Della Lingua Italiana*, 1981). 3. According to the bioregionalist filter, any material in some way incorporates a series of social, cultural, and identity significances of its place, which the bioregionalist decisional process should keep alive by means of adopting specifically local resources.

Mediterranean approach: A mode of designing in the Mediterranean area, which should take into account: the important role of the materials as a decision item within the architectural process, affecting both the shape and the contents of the final building. The climatic context peculiar to these zones. The way of living of the population imprinted in the conviviality and on other principles that could be extracted from the meridian thought: that is, the respect for living and abiotic nature, which refuses free competition and the blind run towards efficiency, besides the trends of economic optimization at short scale, rather than employing traditional methods by substituting as communication means dialogue rather than war and competition. Moreover, the respect for borders as cultural and social exchange, rather than the construction of divisional barriers, the nomadism of ideas and emotions, and finally, a quiet life instead of a chaotic and speedy one (Cassano, 1997).

Nature: 1. "Creative and regulatory physical power conceived of as intermediate cause of phenomena of material world; these phenomena as a whole" (*Oxford Illustrated Dictionary*, 1982). 2. "… a proclaimed system, system of the universe, to be reproduced in the way of the *Homo sapiens'* research" (Oechslin, 1981).

Open space: An open space is a complex braid of social, cultural and economic activities within urban districts, answering to their peculiar logics, just because it is inscribed in that context. It is the simultaneous expression of social realm and physical system, with the microclimate and the physical components of the space.

(Carbone, 2009: 72)

Product: 1. A substance produced during a natural, chemical, or manufacturing process. 2. For Lefebvre, "a work is a work of art, whereas a product is the reproducible result of a mechanical process" (Lefebvre, 1991); 3. "Goods and services obtained as result of the production process" (Roy and Cross, 1979: 134).

Project: 1. The design fact, coming from the Latin *pro-jacere* (throwing before), as a joined action of throwing before so as to achieve a new state and of re-send together, what it is. It is then, in the common sense, a projection onward, in the future, through an action strategy, of a symbol-idea, which is original, unique, as image of a significant structure, and together, a process of its trans-mutations till the establishment of an actual object.

(Ciribini, 1984: 68)

2. All that one proposes to do/the complex of calculations and drawings that estblishes the shape and the characteristics of a construction.

(translated from *Dizionario Garzanti Della Lingua Italiana*, 1981)

Public open space: 1. The urban places in outdoor conditions, where the majority of citizen's interchange occurs. 2. The open spaces in a city are a knotty point of a system of places and paths through which people can enter in symbiosis with both the natural and cultural environment with tourist, ludic, or working goals. 3. Places without any private property, and thus, could be defined as common goods.

Recycling: 1. "To treat or process (used or waste materials) so as to make suitable for reuse" (*Random House Webster's Unabridged Dictionary of American English*, 2015). 2. It is the recovery of useful materials, such as paper, glass, plastic, and metals, from rubbish to make new products, reducing the amount of virgin raw materials needed.

Regeneration actions: 1. Requalification: giving quality to the settlement and areas by upgrading the performance levels, as well as capability to the function's requirement, and the proximity for avoiding obsolescence. 2. Provide revitalization by indicating and creating new use destinations, which will attract people, tourists, and citizens, thus re-energizing places.

Regeneration: 1. A method for providing comfort and livability to people inhabiting a town, as well as tourists and neighbors visiting the places. 2. Open spaces require a number of actions, such as *requalification*, that is, to return quality to the settlement and areas by upgrading performance levels, the adequacy of the functional requirement and proximity while avoiding obsolescence; *to provide revitalization* by indicating and creating new use destinations that will attract people, tourists, and citizens, and thus, return life to the places, and finally *to valorize* the benefits, identity, and the landscape chances and potential as much as possible, by making a soft project with nonaggressive ideas and technologies that can disguise or cancel the identity. 3. It is the right procedure "with the perspective of orienting the territorial and building policies towards a wider sharing and participation, for the administrative institutions' growth" (Schiaffonati and Riva, 2014: 101–2).

Requirement: "What it is required as necessity for running an activity" (Italian standard "Norma UNI 7867"). 2. "The consciousness of a lack and the tension to overcome it" (translated from *Dizionario Garzanti Della Lingua Italiana*, 1981).

Resilience: 1. "The measure of the amount of changing or reversing needed for transforming a system depending on processes and structures which are reinforced each other's, in a different apparatus of processes and structures" (Peterson *et al.*, 1998: 10). 2. "The ability of a system, community, or society exposed to hazards to resist, absorb, accommodate to, and recover from the effects of a hazard in a timely and efficient manner, including through the preservation and restoration of its essential basic structures and functions" (UNISDR, 2009).

Soil: 1. "The ground, upper layer of earth in which plants grow" (*Oxford Illustrated Dictionary*, 1982). 2. "It is the sub-layer indispensible to the development of earth vegetation, and as such it plays a literally fundamental role for the majority of the eco-systems" (Di Fidio, 1990). 3. In order to understand the "common sense" of the word soil, the CRCS (Center for the Research of the Consumption of Soil) define it as

> the open space, according to the modern planning theories, [which] represents the material for shattering the compactness of the historical city and for providing a new shape to its parts, to its settlements principles, to its building types; this is partly true, but with a cultural deep mistake: that of the mixing it with a static vision of the land".

(Vallerini, 2012: 56)

Soil consumption: Any conversion of a nonartificial area into an artificial one.

Source reduction: Designing products to reduce the amount of waste that will later need to be thrown away and to make the resulting waste less toxic.

Sustainable city: 1. A city that does not promote any increase, growth, or any additional constructive activity, but rather only carries, on it, development, improvement, and regeneration. 2. Part of the shared space in between architecture and natural environment.

Sustainable development: 1. "It is a development that meets the needs of the present without compromising the ability of future generations to meet their own needs" (World Commission on Environment and Development, 1987). 2. "A development which allows a lasting satisfaction of the human requirements to be achieved and an improvement of the human life quality...with correct policies and strategies" (Forecaster, 1992). 3. "Over the long period, there is not an interest conflict neither between the poor and the rich countries, nor of individual kind...The concept of sustainable for few does not exist" (Ferguson, 1994).

Sustainable technical solution: Allows the buildings and any other construction to employ less energy, exploit renewables, and guarantee more comfort with fewer costs.

Sustainable technology for urban regeneration: It should be as least artificial and aggressive as possible and it should become more a way of transforming and converting the existing urban objects, architecture products, and facilities in the public open spaces rather than a system for constructing new elements and buildings.

Technique: 1. "The technique is proper of an art and craft; the rules which superintend the art or the craft" (Mercier, 1984). 2. Often techniques have been considered as a negative objective rather than a means, and their influence and necessary contribution have been taken as restrictions to architects' freedom, while at other times the designers have depended on the potential of the same technical innovations. "Throughout the present century architects have made fetishes of technological and scientific concepts out of context and have been disappointed by them when they developed according to the process of technical development, not according to the hopes of architects" (Banham, 1960). 3. "The techniques had engrossed the poetry borders; it had neither built horizons, killed the space, nor imprisoned the poets. In an instant, from the technique progress, dream and poetry come out" (Le Corbusier, 1946). 4. "It is a way of manipulating or transforming the elements of the natural non-human environment in order to control or increase the dominium of man over such environment" (Cresswell, 1984).

Technological practice: The whole of cultural (goals, values, ethical codes, faith in the progress, research will, and creativity), technical (knowledge, capacity and technique, tools, machines, chemistry, live ware, resources, products, and wastes), and organization (economic and industrial activity, professional activity, users, consumers, and syndicates) aspects (see Pacey, 1986).

Technology: 1. "The application of scientific and other knowledge forms arranged for practical aim, by means of…articulated systems involving persons and machines" (Naughton, 1977).2. Technology is the art and study of using technique, that is, the manners of transforming the matter into a product, or a useful object, by means of an anthropic tool, then there cannot be any harm in the use of technology as it is only a means, and therefore, neither good nor bad. 3. It is a bridge between the scientific discovery and the concrete architectural work. 4. "Matter of studies of the transformation processes, taken both during their constructing material, and during their constructing know-how" (Ciribini, 1984). "The doctrinal frame of the transformation processes acting in the matter field as well as in that of the thought" (Ciribini, 1979). 5. "It is the exploration of potential through science" (Vidler, 2012). 6. "The whole of norms over which the art practice, a profession or any other activity are founded, not only manually but also…intellectual" (translated from *Enciclopedia Treccani*). 7. "The study of the various procedures, defining the different architectures. In general, as science and application method, it is founded on the use of particular materials, whose characteristics and internal properties are interrelated with the external requirements" (Dizionario di Architettura e urbanistica (DAU), voice "Tecnologia" by Andrea Silipo (Portoghesi, 2006)).

Transfer stations: They are facilities where municipal solid waste is unloaded from collection vehicles and briefly held while it is reloaded onto larger, long-distance transport vehicles for shipment to landfills or other treatment or disposal facilities.

Waste: 1. "Any substance or object, of which the taker has decided to get rid of, which is included in the categories reported in the Annex A, such as: residual of production; residual of consumption, out of norm products, out of date products, accidentally reversed substances, contaminated or dirty substances, unusable elements, substances become inapt to use, residual of industrial processes, residual of anti-pollution procedures, residual of manufacturing, residual coming from extraction and preparation of raw matters, contaminated substances, any matter, substance or product, whose utilization is forbidden, agricultural and family products which do not have any utility" (Italian Law Decree, 2006). 2. Waste is a human invention. Now we need to spend some effort to "de-invent" it   (http://www.chelseagreen.com/content/zero-waste-sustainability/#sthash.DvLTeGFH.dpuf).

Bibliography

Architecture

AAVV (2003), *Disegnare Paesaggi Costruiti*, Franco Angeli, Milan.

AAVV (2012), *Transitional Spaces*, X-change ed., Vienna.

AAVV (2013), Time 20, in *Architectural Association 20*, Blackmore, Shaftesbury.

AAVV (2014), *Verso la Città Metropolitana di Napoli. Lettura Transdisciplinare*, Luciano ed., Napoli.

Abrami, G. (1987), *Progettazione Ambientale*, Clup, Milan.

Adam, R. (2008), Globalisation and architecture, *Architectural Review*, 1332, February.

Allen, G. (2008), Microcosmi eccellenti, *L'architettura Naturale*, 39, June.

Allen, G. (2008), View, in *L'architettura Naturale*, pp. 40–1, set-dic.

Aravena, A. (2009), Interview, *The Plan*, 33.

Atroshenko, V., Milton Grundy, I. (1991), *Mediterranean Vernacular: A Vanishing Architectural Tradition*, Rizzoli, Milan.

Balducci, A. (1991), *Disegnare il Futuro*, Il Mulino, Bologna.

Banham, R. (1960), The Machine, *Architectural Review*, March.

Banham, R. (1969), *The Architecture of the Well-Tempered Environment*, Chicago Press, Chicago.

Brandi, C. (1977), Teoria del restauro, in L.Vlad Borrelli, J. Raspi Serra, G. Urbani *"Brandi's Lectures"*, Einaudi, Turin.

Buono, M., Masullo, A., Pellegrino, M. (2014), Energie Rinnovabili, in AAVV *"Verso la città metropolitana di Napoli. lettura transdisciplinare"*, Luciano ed., Napoli.

Carbonara, G., Strappa, G. (2013), Enciclopedia di Architettura, UTET Wolters Kluwer, Turin, at the voice "organic pv."

Casoni, G., Fanzini, D., Trocchianesi, R. (2008), *Progetti per lo sviluppo del territorio. Marketing strategico dell'Oltrepo mantovano*, Maggioli editore, Santarcangelo di Romagna.

Congress for the New Urbanism (1993), Charter and principles, Point 6.

Council for European Urbanism (2003), The Charter for EU Urbanism, Stockholm, 6 November, "Action."

Curtis, W.J.R. (2012), Luis Kahn, The space of ideas, *Architectural Review*, 1389, November.

Davey, P. (2000), Sustainable human scale, *Architectural Review*, 1266, August.

De Lucchi, C., Pastenga, D. (2008), Robert and Brenda Vale. Architettura per un mondo finito, *L'architettura Naturale*, 39, June.

Degros, A., De Cleene, M. (2013), *Brussels. Re-discovering its space. Public spaces in the sustainable neighbourhood contacts*. Brussels Capital Region edition, Brussels.

Di Fidio, M. (1990), *Architettura del paesaggio*, Piola ed., Milan.

Finch, P. (2008), View, *Architectural Review*, 1332, February.

Fowler, P.A., Hughes, J., Elias, R.M. (2006), Biocomposites: technology, environmental credentials and market forces, *Journal of the Science of Food and Agriculture*, 86, 1781–9.

Frampton, K. (1987), *Ten points on an architecture of Regionalism. A provisional polemic* in "Center 3 New Regionalism."

Francese, D. (1996), *Architettura bioclimatica*, UTET, Turin.

Francese, D. (2002), *Il benessere negli interventi di recupero edilizio*, Diade, Padova.

Francese, D. (2007), *Architettura e vivibilità: modelli di verifica, principi di biocompatibilità, esempi di opere per il rispetto ambientale*, FrancoAngeli, Milan.

Franck, K.A., Steven, Q. (2013), *Loose Space. Possibility and Diversity in Urban Life*, Routledge, London.

Graves, M. (2012), Architecture and the lost art of drawing, *The New York Times*, 1 September.

Hegel, G.W.F. (1972), *Estetica*, Einaudi, Turin.

Heidegger, M. (1910–1976), *Published Writings*, Vittorio Klostermann, Frankfurt am Main.

Ibelings, H. (1998), *Supermodernism: Architecture in the Age of Globalisation*, NAI Publishers, Rotterdam.

Isola, A. (2003), Al di qua del paesaggio, in AAVV *"Disegnare paesaggi costruiti"*, Franco Angeli, Milan.

Italian Law Decree of the Public Job Ministry (1997), 22 October 1997, in *GURI (Gazzetta Ufficiale della Repubblica Italiana (Official Journal of the Italian Republic))* 30 January 1997, n. 30.

Jacob, S. (2009), James Wines: the outsider, who thinks real culture occurs in car parks, *Architectural Review*, 1354, December.

Jencks, C. (2002), How big is bad, *Architectural Review*, 1266, August.

Jewson, T. (2013), Time 20, in *Architectural Association 20*, Blackmore, Shaftesbury.

Kronenburg, R. (2003), *Portable Architecture*, University of Liverpool.

McHale, J. (1967), The future of the future, *Architectural Design Magazine*, February.

Madanipour, A., Knierbein, S., Degros, A. (2013), *Public Space and the Challenge of Urban Transformation in Europe*, Routledge, London.

Milani, D. (1965), *L'obbedienza non è più una virtù. Documenti del processo di Don Milani*, Firenze, Libreria Editrice Fiorentina, Florence.

Nietzsche, F. (1924), *The complete work*, Volume II "The Will to Power."

Norberg-Schultz, C. (1962), *Genius loci: paesaggio ambiente architettura*, Electa, Milan.

Norberg-Schulz, C. (1968), Less is more, *Architectural Review*, April.

Oechslin, W. (1981), Architettura e Natura, *LOTUS* n. 31, II.

Portoghesi, P., Scarano, R. (2003), *Architettura del Mediterraneo. Conservazione, trasformazione, innovazione*, Gangemi ed., Rome.

Priori, G. (2012), *Intervista ad Aimaro Isola*, in 3M Architetture e città del 3° millennio, n. 4, anno IV.

Sasso, U. (2007), *Riflessi di Bioarchitettura*, Alinea ed., Florence.

Schiaffonati, F., Riva, R. (2014), *Il progetto della residenza sociale*, Maggioli ed., Rimini.

Selicato, F., Cardinale, T. (2012), L'isola di calore nella pianificazione energetica, in *Citta energia*, AAVV, ed., Le Pensieur, Napoli.

Seminario, G. (2009), Intervista a Paolo Portoghesi, *3M Architetture e città del 3° millennio*, n. 0, anno I.

Sgrosso, A. (1979), *Lo spazio rappresentativo dell'architettura*, Lithorapid, Napoli.

Sgrosso, A. (1984), *La struttura e l'immagine*, SEN, Napoli.

Vale, R., Vale, B. (2009). *Time to Eat the Dog: the Real Guide to Sustainable Living*, Thames and Hudson, London.

World Commission on Environment and Development: Our Common Future, Transmitted to the General Assembly as an Annex to document A/42/427 - Development and International Co-operation: Environment, June 1987.

City

AAVV (1999), *Prove aperte di cittadinanza*, GrafiSystem, Modugno, Italy.

Astrade, L., Lutoff, C., Nedjai, R., Philippe, C., Loison, D., Bottollier-Depois, S. (2007), Periurbanisation and natural hazards, *Journal of Alpine Research*, 95, 19–28.

Athens Chart (1931), http://www.icomos.org/en/charters-and-texts/179-articles-en-francais/ressources/charters-and-standards/167-the-athens-charter-for-the-restoration-of-historic-monuments; accessed 31 October 2015.

Bettini, V. (1996), *Elementi di Ecologia Urbana*, Einaudi, Turin.

Bishop, P., Williams, L. (2012), *The Temporary City*, Routledge, London.

Bristow, K. (2014), *Keynote*, in "International conference on peri-urban landscapes: water, food and environmental security", Sydney.

Buccaro, A., Donatone, G., Marselli, M. (2014), Identità storica del territorio e funzioni innovative per la conoscenza e valorizzazione dei beni culturali, in AAVV *"Verso la città metropolitana di Napoli. Lettura transdisciplinare"*, Luciano ed., Napoli.

Buchanan, P. (2012), The big re-think place and aliveness: pattern, play and the planet, *Architectural Review*, August, 86–95.

Carabelli, R., Larribe, S., Bailleul, H. (2011), *R+ O ! Developpement durable et valorisation de l'héritage patrimonial dans la conception des espace publics*, Bonomia university Press, Bologna.

Carbone, I. (2009), Abitare gli spazi aperti, *3M Architetture e città del 3° millennio* n. 0, anno I.

Caruso, G. (2001), *Peri-urbanisation: the situation in Europe. A bibliographical note and survey of studies in the Netherlands, Belgium, Great Britain, Germany, Italy and the Nordic countries*. Report prepared for DATAR, France.

Casagrande, M. (2008), Crossover Architecture and the third generation city, *Epifanio*, 9.

Cimorelli, L., Covelli, C., Cozzolino, L., Della Morte, R., Pianese, D. (2012), *Il recupero della piena efficienza idraulica delle reti urbane di drenaggio mediante il posizionamento ed il dimensionamento ottimali di vasche di laminazione*. Atti del XXXIII Convegno di Idraulica e Costruzioni idrauliche, Brescia, 10–15 September.

De Ambrogio, U. (2009), Come fare un buon progetto partecipato. Una proposta operativa in *"Prospettive sociali e sanitarie"* n° 4, Irs (Istituto per la ricerca sociale), Milan.

De Certeau, M. (1984), *The Practice of Everyday Life*, University of California Press, Berkeley.

De Certeau, M., Clément, G., Casagrande, M. (2013), The Third Infoscape and the recreation of our cities, http://www.ArtisOpenSource.net [AOS].

De Rubertis, R. (2008), *La città mutante*, FrancoAngeli, Milan.

delli Ponti, A. (2012), Soft infrastructures, *Rivista Paesaggio urbano*, 5/6.

Di Fidio, M. (1990), *Architettura del paesaggio*, Piola ed., Milan.

Donadieu, P., Boissien, E. (2001), *Des mots des paysages et des jardins*, ENSP, Versailles, p. 19.

Errington, A. (1994), The Peri-urban Fringe – Europe's Forgotten Rural-Areas, *Journal of Rural Studies*, 10, 367–75.

European Landscape Convention, European Council, article 1.

Evans, G. and Foord, J. (2007), The generation of diversity: mixed use and urban sustainability, in K. Thwaites, S. Porta, O. Ronco, M. Graves eds, *Urban Sustainability, Through Environmental Design*, Routledge Taylor and Francis, London.

Foote Whyte, W. (1991), *Participatory Action Research*, Sage Publications, Thousand Oaks.

Geddes, P. (1970), *Cities in Evolution*, William Sand Norgate, London.

Haydn, F., Temel, R. (2006), *Temporary Urban Spaces: Concepts for the Use of the City Spaces*, Birkhauser, Basel.

Iaconesi, S., Persico, O. (2012), Connect city: real-time observation and interaction for cities using information harvested from social networks, *International Journal of Art, Culture and Design Technologies (IJACDT)*, 2.

Ilardi, M. (2007), *Il tramonto dei non luoghi*, Meltemi, Rome.

Ingersoll, R. (2003), Sprawl-scape: il paesaggio come redenzione, in AAVV *"Disegnare paesaggi costruiti"*, Francoangeli, Milano.

Keith, B. (2014), *Keynote*, in The International conference on peri-urban landscapes: water, food and environmental security, Sydney.

Lambert, A. (2011), The (mis)measurement of periurbanization, *Metropolitics*, 11 May.

Landolfo, R., Manfredi, G., Serino, G., Zuccaro, G. (2014), Rischio Sismico, in AAVV *"Verso la città metropolitana di Napoli. lettura transdisciplinare"*, Luciano ed., Naples.

Le Corbusier (1946), Manière de penser l'urbanisme, Editions de *l'Architecture d'Aujourd'hui*, Paris.

Lefebvre, H. (1991), *The Production of Space*, Blackwell Publishers Ltd, Oxford.

Laozi (1993), *Tao*, Stampa Alternativa ed., Terni (Italy).

Law Decree (2002), n. 42, Codice dei beni culturali e del paesaggio (Code of the cultural goods and landscape), according to the art. 10 of the Law n. 137, of 6 July 2002.

Lucarelli, A. (2011), *Beni Comuni*, Dissensi ed., Viareggio.

Martinotti, G. (1993), *Metropoli. La nuova morfologia sociale della citta*, Il Mulino, Bologna.

Mininni, M. (2012), *Approssimazioni alla città*, Donzelli Editore, Rome.

Moccia, F.D. (2012), *Urbanistica*, CLEAN, Naples.

Mumford, L. (1954), *La cultura delle città*, Edizioni di comunità, Milano

Musil, R. (1957), *Man Without Quality*, Einaudi ed., Milan.

Neuman, M. (2012), Space and place, haste and waste, *Berkeley Planning Journal*, 7, 157–65.

Oswalt, P., Overmeyer, K., Misselwitz, P. (2013), *Urban Catalyst – The power of temporary use*, Dom Publishers, Berlin.

Papastergiadis, N. (2000), *The Turbulence of Migration: Globalisation. De-territorialisation and Hybridity*, Polity Press, Cambridge.

Peregalli, R. (2010), *I luoghi e la polvere*, Bompiani, Milan.

Persico, P (2013), *La città e l'altra città*, Palazzo Bonaretti ed., Novellara, Italy.

Pianese, D., De Vita, P. (2014), Assetto Idrogeologico, in AAVV *"Verso la città metropolitana di Napoli. lettura transdisciplinare"*, Luciano ed., Naples.

Pulselli, R.M., Tiezzi, E. (2008), *Città fuori dal caos. La sostenibilità dei sistemi urbani*, Donzelli ed., Rome.

Purini, F. (2003), in Ilardi, M. (2007), *Il tramonto dei non luoghi*, Meltemi, Rome.

Raffestein, C. (2003), Paysages construits et territorialites, in AAVV *"Disegnare paesaggi costruiti"*, Franco Angeli, Milan.

Ravetz, J. (2014), *URBAN 3.0 Synergistic Pathways for One Planet Cities, Economies and Ecologies in the 21st Century*, Earthscan, Routledge, London.

Register, R. (2006), *Ecocities. Rebuilding Cities in Balance with Nature,* New Society Publisher, Gabriola Island.

Russo, M. (2014), *Urbanistica per una diversa crescita*, Donzelli, Rome.
Salvemini, B. (2006), *Il territorio sghembo*, Edilpuglia, Bari.
Saramago, J. (2000), *La Caverna*, Einaudi, Turin.
Sassen, S. (1991), *The Global City: New York, London, Tokyo*, Princeton University Press, Princeton.
Schon, D. (1983), The reflective practitioner, Basic Books, New York, in Mininni, M. (2012), *Approssimazioni alla città*, Donzelli Editore, Rome, p. 7.
Thwaites, K., Porta, S., Ronco, O., Graves, M. (2007), *Urban Sustainability Through Environmental Design*, Routledge Taylor and Francis, London.
Tschumi, B. (1994), *Event Cities*, MIT Press, Cambridge.
UNCHS Habitat (2001), *Tools to support Participatory Urban Decision Making*, Nairobi.
Vanier, M. (2011), Suburbanization as a Project, *Metropolitics*, 6 April.
Vidler, A. (2012), The great divide: technology versus tradition, *Architectural Review*, 1386.
Vidokle, A. (2003), *Here, There, Elsewhere*, Skive Krabbensholm Books.
Wiegand, D., in Vidokle, A. (2003), *"Here, There, Elsewhere"*, Skive Krabbensholm Books, p. 97.
Zasada, I., Fertner, C., Piorr, A., Nielsen, T.S. (2011), Peri-urbanisation and multifunctional adaptation of agriculture around Copenhagen, *Geografisk Tidsskrift-Danish Journal of Geography*, 111, 59–72.

De-growth

Abramovitz, M. (1979), Economic growth and its discontents, in M.J. Boskin ed., *Economics and Human Welfare. Essays in Honor of Tibor Scitovsky*, Academic Press, New York.
Aries, P. (2005), *Decroissance ou barbarie*, Golias, Villeurbanne.
Arminio, F. (2011), *Terracarne*, Mondadori, Milan.
Bartolini, S. (2010), *Come passare dalla società del ben-avere a quella del ben-essere*, Donzelli, Rome.
Benoits, A. (2007), *Demain, la decroissance! penser l'écologie jusqu'au bout*, Edite, Paris.
Cassano, F. (1997), *Sapere di confine*, in Pluriverso n. 1, RCS ad., Milan.
Daly, HE., Cumberland, J., Costanza, R., Goodland, R., Norgaard, R. (1997), *An Introduction to Ecological Economics,* St. Lucie Press, Delray Beach.
Daly, H. (2001), *Oltre la crescita. L'economia dello sviluppo sostenibile*, Edizioni comunità, Turin.
Duverger, T. (2010), *La Decroissance. En quete d'un capitalisme*, Université de Bordeaux.
Friedman, Y. (2009), *L'architettura di sopravvivenza. Una filosofia della povertà*, Bollati Boringhieri, Turin.
Galimberti, U. (2013), *Speech*, Department of Medicine, University of Naples, 6 March.
Georgescu-Roegen, N. (1995), *La Decroissance. Entropie, écologie, èconomie*, Sang de la Terre ed., Paris.
Gilles, C. (2004), *Manifeste du tiers paysage*, Éditeur Sens et Tonka, France.
Habermas, J. (2006), *The Divided West*, Polity Press, Cambridge.
Heinberg, R. (2011), *The End of Growth*, New Society Publisher, Gabriola Island.
Illich, I. (1974), *La convivialità. Una proposta libertaria per una politica dei limiti allo sviluppo*, Mondadori, Milan.
Kronenburg, R., Lim, J., Wong, Y.C. (2003), *Transportable Environments II*, Routledge, London and New York.
Latouche, S. (1999), *The Degrowth Proposal, Farewell to Growth*, Polity Press, Cambridge.

Latouche, S. (2010), *Come si esce dalla società dei consumi. Corsi e percorsi della Decrescita*, Bollati Boringhieri, Turin.

Latouche, S. (2012), *Discourse*, University of Naples, January.

Masullo, A. (2008), *La sfida del bruco: quando l'economia supera i limiti della biosfera*, Muzzio ed., Rome.

Masullo, A. (2013), *Qualità verso Quantità. Dalla decrescita a una nuova economia*, Lit ed., Rome.

Moretti, G. (2005), La rete Bioregionale Italiana, *Ecologia profonda*, November.

Rebb, T. (2005), http:// www.ecologiaprofonda.com; accessed 31 October 2015.

Sale, K. (2000), *Dwellers in the Land: The Bioregional Vision*, University of Georgia Press, Athens.

Sassen, S. (1997), *Le città nell'economia globale*, Il mulino, Bologna.

Sassoli, E. (2008), *Non solo shopping. Usi sociali dei luoghi di consumo*, Le Lettere ed., Florence.

Secchi, B. (2014), Dialogo oltre la crescita, in Russo, M., *Urbanistica per una Diversa Crescita*, Donzelli, Rome.

Soleri, P. (2009), L'urbanizzazione frugale come corrente alternativa al materialismo, all'iperconsumismo e allo sprawl urbano, *Le carré bleu*, 1.

Tiezzi, E. (2005), *Tempi storici, tempi biologici*, Garzanti ed., Milan.

Voltaire (1995), Senso comune, in Segre, B., *Dizionario filosofico*, BIT, Milan, p. 288.

Environmental issues

AAVV Ambiente Italia (2003), *Indicatori comuni Europei: verso un Profilo di Sostenibilità Locale*, Ancora Arti Grafiche, Milan.

AAVV (2012), *Citta energia*, Le Pensieur ed., Naples.

Alexander, D. (1996), Bioregionalism: the need for a firmer theoretical foundation, *Journal of Ecosophy*, 3.

Aristotle, *Metaphysics*, translated by Hardie, R. P. and Gaye, R. K. http://arpast.org/newsevents/articles/article47.pdf; accessed 20 December 2015.

Assunto, R. (1980), Paesaggio-Ambiente-Territorio. Un tentativo di Precisazione concettuale, *Bollettino CISA*, XVIII.

Bauman, Z. (2001), *Voglia di comunità*, LaTerza, Bari.

Bauman, Z. (2013), Dalla competizione alla cooperazione, in A. Masullo ed., *Qualità verso quantità*, Orme, Rome.

Berg, P. (1984), *Welcome speech from the First Bioregional Congress of North America*.

Berg, P., Dasmann, R. (1977), Reinhabiting California, *Ecologist*, 7, 399–401.

Bettini, V., Falqui, E., Alberti, M. (1984), *Il bilancio di impatto ambientale*, Clup, Milano.

Bevilacqua, P. (1996), *Tra natura e storia. Ambiente, economie, risorse in Italia,* Donzelli, Rome.

Borghini, T., Tatavitto, M. (2012), Lo scenario energetico territorializzato della Provinciadi Prato, in AAVV "*Citta energia*", Le Pensieur ed., Naples.

Borja, J., Castells, M. (1997), *Local and Global. Management of Cities in the Information Age*, Earthscan, Routledge, London.

Carson, R. (2000), *Silent Spring*, Penguin Books, London.

Ciribini, G. (1990), *La valutazione di impatto ambientale*, Alinea, Florence.

Ciribini, G., Gasparotti, R. (1990), Significato e riferimenti dell'impatto ambientale, in Ciribini G. *La valutazione di impatto ambientale*, Alinea, Florence, p. 49.

Cobb J.B. Jr., (2001), Deep ecology and process thought, *Process Studies*, 30, 112–31.

Corradi L. (2008), *Salute e Ambiente*, Carocci, Rome.

Daly, H.E. (1990), Towards some operational principles of sustainable development, *Ecological Economics* n. 2.

Devall, W. and Sessions G. (1985), *Deep Ecology: Living As if Nature Mattered*, Gibbs M. Smith, Inc., Salt Lake City.

Devoto, G., Oli, G.C. (1981), *Dizionario della lingua italiana, voice: Benessere*, Le Monnier, ed., Firenze.

Enciclopedia Treccani, http://www.treccani.it/enciclopedia; accessed 31 October 2015.

Ferguson, E.T. (1994), *Sustainability and energy policy*, in World Renewable Conference Proceedings, Reading ed., London.

Forecaster, F. (1992), Il bello, il buono, l'utile dello sviluppo sostenibile, in *"Prometeo"*, 39, September.

Gardi C., Montanarella L., Panagos P. (2013), Strategie europee per il contenimento dei consumi di suolo, in *Il progetto sostenibile* n° 33, p. 25.

Gardner G., Prugh T. (2008), *Seeding the Sustainable Economy*, in State of the World, W.W. Norton & Company, New York.

Geddes, P. (2012), *The Evergreen: a Northern Seasonal*, Forgotten Books, London. (Original work published 1895.)

Giddens, A. (2005), Le politiche del cambiamento climatico, in *Economia e Ambiente*, ed., Missionaria Italiana, EMI, Bologna.

Heidegger, M. (2001), *Saggi e Discorsi*, Mursia Editore, Milan.

Hui, C. (2006), Carrying capacity, population equilibrium, and environment's maximal load, *Ecological Modelling*, 192, 317–20.

Italian Law Decree (2006), n. 152, 3 April 2006, Norme in materia ambientale (Standards for the Environment) published in *GURI (Gazetta Ufficiale della Repubblica Italiana (Official Journal of the Italian Republic))* n. 88 14 April 2006.

Levi-Montalcini, R. (2009), Tempo di mutamenti, *Le carré bleu*, n° 1.

Marotta, P., Schilleci, F. (2012), Il territorio e l'uso delle energie rinnovabili nella città, in AAVV *"Citta energia"*, Le Pensieur ed., Naples.

Mattoscio, N. (1999), Globalizzazione, domanda effettiva e occupazione, *Global & Local*, Volume I.

Mazzeo, A., Fasolino, A.R. (2014), Informatica, in AAVV *"Verso la città metropolitana di Napoli. lettura transdisciplinare"*, Luciano ed., Naples.

Millennium Ecosystem Assessment, in Elisabetta Peccol (2013), Infrastruttura verde e consumo di suolo, *Il progetto sostenibile* n. 33, 42–3.

Munafò, M. (2013), La misurazione del consumo di suolo a scala nazionale, *Il progetto sostenibile*, n. 33.

Okakura, K. (1906), *The Book of Tea*, Watchmaker Publishing, Seaside.

Peccol, E (2013), Infrastruttura verde e consumo di suolo, *Il progetto sostenibile* n. 33, pp. 42–3.

Peterson, G., Allen, C.R., Holling, C.S. (1998), Ecological resilience, biodiversity and scale, *Ecosystem*, 1.

Pickard, R. (2001), *Management of Historic Centres*, Taylor and Francis, Oxford.

Rees, W.E. (1992), Ecological footprints and appropriated carrying capacity, *Environment and Urbanization*, 4, 121–30.

Rees, W.E. (2011), *The human nature of unsustainability*, Post carbon institute ed., USA.

Rees, W.E., Wackernagel, M. (1994), *Ecological Footprints and Appropriated Carrying Capacity: Measuring the Natural Capital Requirements of the Human Economy*, Island Press, Washington DC.

Santarelli, E., Figini, P. (2004), *Does globalization reduce poverty? Some empirical evidence for the developing countries*, Alma mater digital library, Bologna.

Schumacher, E.F.F. (1973), *Small is Beautiful; A Study of Economics as if People Mattered*, Blonde and Briggs Ltd, London.

Semerari, G. (2003), *Da Hiram a Federico II* in P. Portoghesi e R. Scarano Architettura del Mediterraneo. Conservazione, trasformazione, innovazione, Gangemi ed., Rome, p. 109.

Stevenson, F. (2013), *A bioregional approach to climate change design*, in ARC 307, 327, 377, Environment and Technology 5.

Tanese, P. (1999), Premessa, in AAVV *"Prove aperte di cittadinanza"*, GrafiSystem, Modugno, Italy.

Texas Monthly (2008), http://www.texasmonthly.com/articles/pliny-fisk-iii-gail-vittori/; accessed 31 October 2015.

Tozzi, M. (2013), Prefazione, in Andrea Masullo *"Qualità verso quantità"*, Orme, Lit ed., Rome.

UNISDR (2009), Terminology on Disaster Risk Reduction.

Vale, B., Vale, R., Mithraratne, N. (2007), *Sustainable Living: the Role of Whole Life Costs and Values*, Butterworth-Heinemann, Elsevier, Atlanta.

Vallerini, L. (2012), Suolo, risorsa di paesaggio, *Architettura del paesaggio*, n. 26, maggio settembre.

Van Newkirk, A. (2009*)*, Bioregions: towards bio regional strategy for human cultures, *Environmental Conservation*, 2, 108.

Vanoli, L. (2012), Fonti energetiche e inquinamento urbano, in AAVV *"Citta energia"*, Le Pensieur ed., Naples.

Vercelloni, V. (1992), *Ecologia degli insediamenti umani*, Jaca Book, Milan.

Wackernagel, M. (2005), Il nostro pianeta si sta esaurendo, in AAVV, *Economia e ambiente*, EMI ed., Bologna.

Wackernagel, M., in Antony Giddens (2005), *Le politiche del cambiamento climatico*, *"terza via" AAVV Economia e Ambiente, Ed Missionaria Italiana*, EMI, Bologna, pages 93–94.

Wackernagel, M., Rees, W.E. (1996), Our ecological footprint – Reducing human impact on the Earth, *Environment and Urbanization*, 8.

WWF (2006), *The Living Planet Report*.

Recycle, reuse, reduce

Bronari C. *et al.* (2012), *Tecnologie per il riciclo/riciclaggio sostenibile dei rifiuti*, EAI Bimestrale Enea SPECIALE I.

Callister, W.D. (2007), *Materials Science and Engineering: An Introduction,* John Wiley and Sons, Hoboken.

Congress for the New Urbanism (1993), *Charter and Principles*, Point 6.

Connett, P. (2013), *The Zero Waste Solution: Un-trashing the Planet. One Community at a Time,* Chelsea Green, White River Junction.

Connett P. (2014), *Zero waste: a concrete step towards sustainability.* http://www.chelseagreen.com/content/zero-waste-sustainability/#sthash.DvLTeGFH.dpuf; accessed 31 October 2015.

Crary, J. (2013), Time 20, in *"Architectural Association 20"*, Blackmore, Shaftesbury.

Curzio, Q. (1999), Globalizzazione: profili economici, *Il risparmio*, 2, 202.

Dizionario Garzanti della Lingua Italiana (1981), Garzanti ed., Milano.

Encyclopedia Britannica (2015), Recycling, in: http://www.britannica.com/science/recycling; accessed 31 October 2015.

European Directive (2006) 2008/98/CE art.6. From the waste to the "secondary prime matter."

Fedorchenko, M. (2013), Time 20, in *"Architectural Association 20"*, Blackmore, Shaftesbury.

Frediani, G. (2012), Eminonu, comments on an urban project, in AAVV *"Transitional Spaces"*, X-change ed., Vienna, p. 12.

Italian Law Decree (2006), 152, Ulteriori disposizioni correttive ed integrative del decreto legislativo 3 aprile 2006, n. 152, recante norme in materia ambientale.

Italian Law Decree (2010), 205, as application of the European Directory 2008/98/Ce, and published on the GURI (Gazzetta Ufficiale della Repubblica Italiana- Official Journal of the Italian Republic) on 10 December 2010, n. 288.

Lynch K. (1990), *Waste Away*, Sierra Club Book, San Francisco.

Mak, J. (2013), Time 20, in *"Architectural Association 20"*, Blackmore, Shaftesbury.

Oxford Illustrated Dictionary, (1982), Coulson, J., Carr, C.T., Hutchinson, L., Eagle, D., eds, Clarendon Press, Oxford.

Passaro A. (1996), *Costruire e dismettere*, Arte tipografica, Naples.

Peck, A. (2014), *What is the difference between reuse, reduce and recycle?* http://homeguides.sfgate.com; accessed 20 December 2015.

Random House Webster's Unabridged Dictionary of American English (2015), Random House, New York.

U.S. Environmental Protection Agency (2010), voices: Reduce and Reuse, Recycling, What You Can Do on the Go, Municipal Solid Waste Generation, EPA; December.

Vale, R., Vale B. (1975), *The Autonomous House*, Universe Books, New York.

Wieman, B. (2014), *What is reduce, reuse and recycle?* http://homeguides.sfgate.com; accessed 20 December 2015.

Yildiz, D. (2012), Conceptual interventions for creating livable environments on the two sides of the Galata bridge, in AAVV *"Transitional Spaces"*, X-change ed., Vienna, p. 26.

Technology

AAVV (1989), *Tecnologie del recupero edilizio*, UTET, Turin.

Banham, R. (1960), *Theory and Design in the First Machine Age*, The Architectural Press, London.

Ciribini, G. (1979), *Introduzione alla tecnologia del Design*, Franco Angeli, Milan.

Ciribini, G. (1984), *Tecnologia e Progetto*, Celid, Turin.

Cresswell, R. (1984), voice "tecnica", Enciclopedia Einaudi, Turin.

De Joanna P., Francese D., Passaro A. (2012), *Sustainable Mediterranean Construction*, Franco Angeli, Milan.

Galimberti, U. (2004), *Psiche e Techne*, Feltrinelli, Milan.

Italian Standard "Norma UNI/CEE 0050: Classi di esigenze e relative definizioni" (Requirements classes and their definitions), in Zaffagnini, M. (1981), *Progettare nel processo edilizio*, L. Parma, ed., Bologna, p. 121.

Italian Standard "Norma UNI 7867: Scomposizione del sistema edilizio" (Composition of the building system), in Zaffagnini, M. (1981), *Progettare nel processo edilizio*, L. Parma, ed., Bologna, p. 121.

Mercier, A. (1984), Tecnica e tecnologia. Tra natura e cultura, *Nuova civiltà delle macchine*, 2, 2.

Naughton, J. (1977), Introduction: technology and human values, in AAVV, *Living with Technology: A Foundation Course*, The Open University Press, Milton Keynes.

Pacey, A. (1986), *The Culture of Technology*, Editori riuniti, Roma.

Portoghesi, P. (2006), *Dizionario Enciclopedico di Architettura e urbanistica (DAU)*, Gangemi editore, Rome.

Roy R., Cross, N. (1979), *La tecnologia e i suoi effetti sull'economia e sui rapporti sociali*, Mondadori, Milan.

Sorlini, S., Mentore, V. (2009), L'evoluzione delle tecnologie appropriate, in *Tecnologie Appropriate Nella Cooperazione Internazionale Allo Sviluppo*, 18 December, Brescia University, Brescia.

Index

Printed and bound by CPI Group (UK) Ltd, Croydon, CR0 4YY

22/10/2024

01777613-0006